Sustainable Development Goals Series

The **Sustainable Development Goals Series** is Springer Nature's inaugural cross-imprint book series that addresses and supports the United Nations' seventeen Sustainable Development Goals. The series fosters comprehensive research focused on these global targets and endeavours to address some of society's greatest grand challenges. The SDGs are inherently multidisciplinary, and they bring people working across different fields together and working towards a common goal. In this spirit, the Sustainable Development Goals series is the first at Springer Nature to publish books under both the Springer and Palgrave Macmillan imprints, bringing the strengths of our imprints together.

The Sustainable Development Goals Series is organized into eighteen subseries: one subseries based around each of the seventeen respective Sustainable Development Goals, and an eighteenth subseries, "Connecting the Goals," which serves as a home for volumes addressing multiple goals or studying the SDGs as a whole. Each subseries is guided by an expert Subseries Advisor with years or decades of experience studying and addressing core components of their respective Goal.

The SDG Series has a remit as broad as the SDGs themselves, and contributions are welcome from scientists, academics, policymakers, and researchers working in fields related to any of the seventeen goals. If you are interested in contributing a monograph or curated volume to the series, please contact the Publishers: Zachary Romano [Springer; zachary.romano@springer.com] and Rachael Ballard [Palgrave Macmillan; rachael.ballard@palgrave.com].

Andrew Simmons
Andree-Ann Simmons

Building Institutions for Resilience

Combatting Climate Change

Springer

Andrew Simmons
Sustainable Development Consultancy
JEMS Environmental Management Services
Enfield, Middlesex, UK

Andree-Ann Simmons
Sustainable Development Consultancy
JEMS Environmental Management Services
Enfield, Middlesex, UK

ISSN 2523-3084 ISSN 2523-3092 (electronic)
Sustainable Development Goals Series
ISBN 978-3-031-28880-7 ISBN 978-3-031-28881-4 (eBook)
https://doi.org/10.1007/978-3-031-28881-4

This Springer imprint is published by the registered company Springer Nature Switzerland AG
The registered company address is: Gewerbestrasse 11, 6330 Cham, Switzerland

Foreword

During our doctoral research in climate change mitigation and adaptation, Dr Andrew Simmons and I met at the Institute for Energy and Sustainable Development, De Montfort University, Leicester. While I was keen on interrogating the vulnerability of critical oil and gas infrastructure in the Niger Delta to climate change, Andrew dedicated his thesis to investigating young people in Small Island Developing States (SIDs) exposure to climate change impacts. His work turned out to be a vast and novel contribution to the body of knowledge. It remains a reference material for practitioners and experts looking to solve the impact of climate change on the young global population.

Nonetheless, climate change has continued to impact the globe, evidenced by wildfires, floods, storms and other hazards increasing in frequency and severity. The chief weakness arising from the conversation about the devastating impacts is relevant agencies' lack of institutional resilience and adaptive capacities to adapt and mitigate climate change-related disasters. Hence, "Building Institutions for Resilience: Combatting Climate change" is a timely and intervening piece of strategic contribution by Andrew's combined practical experiences and background in climate resilience literature.

The contribution in this book is arranged in independent chapters focusing on building the capacity of national institutions as a strategy for enhancing the resilience of people and communities towards climate change impacts and adaptation. This is scripted using experiences from St Vincent and the Grenadines and Sierra Leone's sustainable development activities in SIDS and post-conflict less-developed countries (LDC), respectively. The book includes case studies from international capacity-building with extensive experiences from the Commonwealth Youth Programme, Caribbean Youth Network and Uganda, as well as lessons learned from the field to address questions that could be raised while trying to implement the approaches and methodologies narrated in this book.

No doubt, this book adds to the series dedicated to climate impact and resilience in developing countries where the effects of climate change are mostly felt. Without fear of contradiction, I am confident that this book is an essential read and reference material for climate change activists, policymakers, practitioners, students, researchers and enthusiasts who deliberate on building institutions for resilience and combatting climate change globally.

Senior Consultant – Climate Change Adaptation,
Net Zero, and Sustainability,
Transport Consulting and Advisory
Atkins Global – A Member of SNC-Lavalin,
London, UK

Justin Udie

Abstract

Climate change is the single most devastating problem affecting every nation that can impede the world from achieving the Sustainable Development Goals. Although most nations signed up to the Paris Climate Agreement in 2015, i.e. keeping the rise in global temperatures well below 2 °C this century and pursuing efforts to keep it under 1.5 °C, under all the emissions scenarios considered by the scientists, research shows both targets will be unmet this century unless concerted actions on the reductions in the emission of carbon are implemented globally. The Paris Climate Change Agreement is a binding international treaty on climate change adopted by 196 parties at COP21 in Paris on December 12, 2015, and entered into force on November 4, 2016. There are enormous challenges regarding implementing the Paris Agreement on Climate Change and the UN Sustainable Development Goals (SDGs). Notably, most developing countries lack the institutional capacity to implement the Paris declaration and the Sustainable Development Goals.

Capacity-building is essential in building the population's resilience to adapt successfully to climate change's impact. Therefore, it is an integral component of climate change adaptation but is severely deficient in developing countries, where financial resources and viable human capital are limited. The book provides insights and learning (strategies and methodologies) on the unique contributions of institutions to building people's capacities to adapt to the impact of climate change and other development calamities in developing countries.

The book offers new perceptions to policymakers, community leaders/practitioners and academics working on climate change adaptation and resilience-building practices and strategies in vulnerable communities, including Small Island Developing States (SIDS) and Least Developing Countries (LDC), focusing on innovative techniques to build the institutional capacity utilizing participatory human capital top-down and bottom-up approaches. Presenting case studies on actions, processes and practices used to enhance the capacity of the institutions involved in climate change adaptation and other development programs, the book reviews several conceptual frameworks, models and theories on adaptation capacity-building for climate change. A new conceptual framework for capacity-building of institutions for climate change adaptation entitled the "Climate Change Adaptation Framework for Capacity-Building in Developing Countries" has been created, building on the best practices highlighted in the cases. The book also makes recommendations to enhance the institutional capacities of organizations in developing countries to effectively build the resilience of people and their communities to respond to climate change.

Using data collection methodologies such as interviews, questionnaires, focus group discussions and case studies, the book assesses the relevance and effectiveness of the capacity-building strategic actions implemented by non-governmental organizations (NGOs), inter-governmental organizations (IGO) and governmental institutions at national, regional and international levels to develop the capacity of institutions to deliver climate change adaptation and sustainable development programs in developing countries.

Contents

Acronyms

ACT	Action on Climate Today
AfDB	African Development Bank
ALP	Accelerated Learning Program
APF	Adaptation Policy Framework
BSC	Business Support Centre
CANARI	Caribbean Natural Resources Institute
CAPS	Career Advising and Placement Services centres
CBO	Community-Based Organizations
CCA	Climate Change Adaptation
CCA	Caribbean Conservation Association
CCCCC	Caribbean Community Centre for Climate Change
CCESD	Climate Change Education for Sustainable Development
CDB	Caribbean Development Bank
CDD	Community Development Division
CDD	-Community-Driven Development
CDO	Community Development Officer
CIDA	Canadian International Development Agency
COM	Committee of Management
COP	Conference of Parties (to UNFCCC)
CPA	Country Poverty Assessment
CPAP	Country Programme Action Plan
CPD	Country Programme Document
CRCAE	Caribbean Regional Council for Adult Education
CSO	Civil Society Organization
CWSA	Central Water and Sewage Authority
CYEN	Caribbean Youth Environment Network
CYMM	Commonwealth Youth Ministers Meeting
CYP	Commonwealth Youth Programme
CVA	Community-based Vulnerability and Adaptation
DfID	Department for International Development
EPC	Environmental Protection Clubs
FGD	Focus Group Discussion
GCF	Green Climate Fund (established by UNFCCC)
GEF	Global Environment Facility
GEF/SGP	Global Environment Facility Small Grants Programme
GIP	Graduates Internship Programme
GYDI	Global Youth Development Index

GYNED-SL	Global Youth Network for Empowerment and Development Sierra Leone
ILO	International Labour Organization
INDC	Intended National Determined Contribution
INGO	International Non-Governmental Organization
IPCC	International Panel for Climate change
ITC	Information and Telecommunication
IT	Information Technology
IYDEN	International Youth Development and Environment Network
JEMS	Junction Enhams McCarthy and Surrounding Areas
LCD	Least Developing States
MCMS	Micro Credit Management System
MER	Monitoring, Evaluation Reporting
MoYA	Ministry of Youth Affairs
MOU	Memorandum of Understanding
NAME	National Association of Mass Organization
NAP	National Adaptation Plan
NAYCOM-SL	National Youth Commission of Sierra Leone
NEMO	National Emergency Management Organisation
NGO	Non-Governmental Organization
NRC	Norwegian Refugee Council
NYC	National Youth Council
NYEEF	National Youth Economic Empowerment Fund
NYP	National Youth Policy
OAS	Organization of American States
OECS	Organisation for the Eastern Caribbean States
OECD	Organisation for Economic Corporation and Development
PAC	Project Appraisal Committee
PI	Partner Institution
PPP	Public-Private Partnership
PLA	Participatory Learning Approach
PRA	Participatory Rural Appraisal
PRSP	Poverty Reduction Strategy Plan
RAY	Recovery Acholi Youth
SABI	Sierra Leone Agricultural Business Initiative
SBI	Subsidiary Body for Implementation
SDGs	Sustainable Development Goals
SEDI	South East Development Inc
SIDS	Small Island Developing States
SLANGO	Sierra Leone Association of Non-Governmental Organizations
TOT	Training of Trainers
TVET	Technical and Vocational Education and Training
UN	United Nations
UNCED	United Nations Conference on Environment and Development
UNDP	United Nations Development Programme

UNEP	United Nations Environment Programmes
UNESCO	United Nations Educational, Scientific and Cultural Organization
UNFCCC	United Nations Framework Convention on Climate Change
UNICEF	United Nations International Children's Emergency Fund
USAID	United States Agency for International Development
VSO	Volunteer Service Overseas
YEP	Youth Education Pack
YEEP	Youth Employment and Empowerment Project
YES	Youth Environment and Service Clubs

1 Introduction

Abstract

The chapter starts broadly with climate change and its impact on LDCs, SIDSs and developing countries in Africa, the Caribbean and the Pacific. It shows climate change as the single most devastating threat facing humanity, potentially hindering the globe from achieving Sustainable Development Goals. It discusses SIDS and LDC's challenges in developing their institutional capacity to respond to climate change. It also provides insights into relevant capacity development models and practices in institutions that participated in the case studies selected to be included in this book.

The chapter discusses the book's purpose, which is to provide insights into the unique contributions of institutions to building the capacities (human) of institutions to adapt to climate change and other development calamities. The book explores relevant development models, conceptual frameworks and strategic programmes from the past as a basis for developing a new conceptual framework, extracting lessons learned and best practices and making recommendations for the future.

Keywords

Climate change adaptation · Human capacity · Sustainable development · Small island developing state · Least developing states · Institutional capacity · Threat · Development calamities · Sustainable Development Goals · Small island developing states

1.1 Challenges Faced by Developing Countries to Achieve Sustainable Development

This chapter begins with climate change and its impact on LDCs, SIDS and developing countries in Africa, the Caribbean and the Pacific. It shows climate change as the single most devastating threat facing humanity, potentially hindering the world from achieving Sustainable Development Goals. It discusses SIDS and LDC's challenges in developing their institutional capacity to respond to climate change and provides insights into relevant capacity development models and practices in institutions participating in the study.

The chapter discusses the book's purpose, which is to provide insights into the unique contributions of institutions to building the capacities (human) of institutions to adapt to climate change and other development calamities. The book explores relevant development models, conceptual frameworks and strategic programmes from the past as a basis for developing new conceptual frameworks, extracting lessons learned and best

A. Simmons, A.-A. Simmons, *Building Institutions for Resilience*, Sustainable Development Goals Series, https://doi.org/10.1007/978-3-031-28881-4_1

practices and making recommendations for new and future developments.

The ground-breaking IPCC report (BBC, 2021a; IPCC, 2022) warns of the increasing presence of extreme heatwaves, droughts and flooding and a critical temperature limit being broken in just over a decade. The UN Secretary-General António Guterres stated that climate change "is a code red for humanity" (United Nations, 2021). The report presented an assessment of the state of the planet and its future, delivered by the UN's Intergovernmental Panel on Climate Change. The report stated that human influence on the warming of the atmosphere, oceans and land is undeniable. The last central assessment review on the science of climate change was in 2013. The release came a few months before the climate summit, COP26, in Glasgow, Scotland, UK, in November 2021.

The report came during mass evacuations as wildfires spread in Greece and the Western USA, millions of people at risk of flooding as seen in China and Germany, exposure to hurricanes in the Caribbean and other weather events and other development calamities as observed in the USA and the Caribbean. These climatic events occurred during the COVID-19 pandemic, which infected millions of people and accounted for over one million deaths worldwide.

Critical summary of the IPCC (2021) report
The global surface temperature was 1.09 C higher between 2011 and 2020 than between 1850 and 1900.
The past 5 years have been the hottest on record since 1850.
The recent rate of sea level rise has nearly tripled compared with 1901–1971.
Human influence is "very likely" (90%) the main driver of the global retreat of glaciers since the 1990s and the decrease in Arctic sea ice.
It is "virtually certain" that hot extremes, including heatwaves, have become more frequent and intense since the 1950s, while cold events have become less frequent.

The IPCC report (IPCC, 2022) states that the current rate of warming experienced globally has contributed to changes in the global support systems. These changes are irreversible in terms of the timescales of centuries. For example, the oceans will warm and become more acidic, mountains and polar glaciers will continue melting for decades or centuries, and forests and ecosystems will experience significant difficulties returning to their past healthy existence.

The IPCC report is followed by a new statement on the world's trees, claiming that at least 30% of the 60,000 tree species faced extinction (BBC, 2021b). This is mainly due to human actions. Approximately 142 tree species have already become extinct from the wild, while 442 are about to be extinct. The report states that the biggest threat to trees is forest clearance for crop production (impacting 29% of species), logging (27%), support for livestock grazing or farming (14%), clearance for development (13%) and fire (13%). The report concludes that climate change, extreme weather and sea level rise continue to be growing threats to trees and ecosystems and conservation actions provide much hope for trees' future. These factors will continue to prolong developing countries' vulnerability to climate change. The IPCC report (2022) demonstrates the intense focus on the interactions among the combined climate systems, ecosystems/biodiversity and human society. It shows that these interactions are the core of the emerging risks from climate change, ecosystem degradation and biodiversity loss. At the same time, they offer opportunities for future sustainability and preservation. The report states that risk can be evolved from the interactions among climate-related hazards and the exposure and vulnerability of affected human and ecological systems. The introduction of risk by human responses to climate change is a new aspect to be considered in the climate risk concept (IPCC, 2022).

Vulnerability is viewed as the function of a system or organization's exposure, sensitivity and adaptive capacity to the impact of climate change (IPCC, 2014a, b, c) and is a component of risk (IPCC, 2022). These concepts are considered significant vulnerability determinants in SIDS and LDCs. According to the literature, exposure, sensitivity and adaptive capacity are not categorized as impact. Vulnerability is widely understood to differ within communities and across societies, regions and countries, also changing through time. However, they are viewed as factors or processes that can increase or reduce the impact of climate change pressures/stressors on communities/countries (IPCC, IPCC, 2014a, b, c; 2022).

Adaptation and resilience are key concepts related to adapting to climate change. According to the IPCC report (2022), adaptation is crucial in reducing the exposure and vulnerability associated with climate change. Adaptation is defined in human systems "as the process of adjustment to actual or expected climate and its effects to moderate harm or take advantage of beneficial opportunities" (IPCC, 2022 p6). Adaptation is viewed as the process of adjustment to the actual climate and its impact. Adaptation in ecological systems includes independent adjustments through biological and evolutionary processes. Therefore, adaptation can be anticipatory or reactive, as well as incremental and/or transformational.

Adaptation is often structured around resilience. It is a process of "bouncing back and returning" to a previous state after a disturbance or disruption (IPCC, 2022). The concept illustrates the ability to maintain essential function, identity and structure, but also the capacity for transformation and revolutionary change. The IPCC report acknowledges the significance of diverse forms of knowledge, such as scientific, Indigenous and local knowledge, as a medium to enhance understanding and evaluate adaptation processes and actions to reduce risks from human-induced climate change.

Climate justice was viewed as an essential development concept in the fight to adapt to climate change according to the IPCC report (2022; p6) and is defined as the capacity of social systems, economic systems and ecosystems to cope with a hazardous event or trend or disturbance, responding or reorganizing in ways that maintain their essential function, identity and structure as well as biodiversity in case of ecosystems while also maintaining the capacity for adaptation, learning and transformation.

The report states that resilience is viewed as a positive attribute when it maintains such a capacity for adaptation, learning and/or transformation. Climate justice typically comprises three fundamental principles:

- Distributive justice: the allocation of burdens and benefits among individuals, nations and generations
- Procedural justice: who decides and participates in decision-making
- Recognition: basic respect, robust engagement and fair consideration of diverse cultures and perspectives

As climate change continues to impact human development, the UN introduced the Sustainable Development Goals (SDGs) in 2015 to provide a model to address the global challenges people are encountering. SDGs include poverty, climate change, environmental degradation, peace and justice and devised a set of strategic actions to enable people across the globe to resolve these development challenges. The mission statement of the SDGs is "a blueprint for achieving a better and sustainable future for all people and the world by 2030" (www.un.org). These goals include:

1. No poverty
2. Zero hunger
3. Good health and well-being
4. Quality education
5. Gender equity
6. Clean water and sanitation
7. Affordable and clean energy
8. Decent work and economic growth
9. Industry, innovation and infrastructure
10. Reduced inequalities
11. Sustainable cities and communities
12. Responsible consumption and production
13. Climate action
14. Life below water
15. Life on land
16. Peace, justice and strong institutions
17. Partnership for the goals (UNDP, 2015)

Although the SDGs are reasonable development goals, developing countries will encounter severe difficulties achieving these developing goals due to financial deficiencies, low institutional capacity and poor and short-term planning and development challenges.

Research shows that climate change is the single most devastating problem affecting every nation (IPCC, 2014a, b, c; UNDP, 2015; Nurse et al., 2014; IPCC, 2022; UN, 2021). It is of high confidence that it possesses the potential to derail the world from achieving Sustainable Development Goals. Although most nations signed up to the Paris Climate Change Agreement in 2015, i.e. keeping the rise in global temperatures well below 2 °C this century and pursuing efforts to keep it under 1.5 °C, under all the emissions scenarios considered by the scientists, both targets will be broken this century. This is unless action on carbon emission reductions is implemented globally. The Paris Climate Change Agreement is a legally binding international treaty on climate change adopted by 196 parties at COP21 in Paris on December 12, 2015, and entered into force on November 4, 2016. However, there are enormous challenges regarding implementing the Paris Climate Change Agreement, bearing in mind the numerous social, economic and development challenges countries face regarding the COVID-19 pandemic and other development calamities (United Nations, 2021; UNFCCC, 2021).

In situations where the political will exists among developing and developed countries, they cannot provide the necessary financial and technological resources to support the implementation of the required adaptation and mitigation actions to reduce the impact of greenhouse gas emissions and climate change. According to the Paris COP21 Climate Change Agreement, developed countries are responsible for providing financial assistance to countries more vulnerable to climate change. The treaty encouraged developed countries to provide voluntary financial contributions to nations to finance their national plans. However, despite countries' commitment to providing financial resources and technical assistance to enhance poorer countries' institutional capacity to implement their obligations to the agreement, these resources are not forthcoming to support climate change adaptation and mitigation initiatives in developing countries. Developed countries failed to provide the necessary new finances to help the 2015 Paris Climate Change Agreement (United Nations, 2021; UNFCCC, 2021). Even when financial resources are available to implement such programmes, most developing countries lack the intuitional capacity, strategic intent and expertise to successfully build their resilience to adapt to the impacts of climate change and development calamities.

1.1.1 Building the Capacity of Institutions to Support Sustainable Development

Capacity-building is essential in the march towards building the population's resilience to adapt successfully to climate change's impact. It is critical to note that institutional capacity-building is a severe deficiency in LDCs, SIDS and other developing countries, where financial resources and viable human capital are limited. Therefore, efforts to increase the capacity of governments and stakeholders' institutions to create adaptive capacity should be based on an awareness of the existing capacity levels within the institutions and/or at the state.

Overall, capacity-building is classified as endeavours and/or actions to build individuals' and organizations' understanding, skills and knowledge base to enable them to achieve their development goals (Bloomfield et al., 2017). It is also described by Gordon and Chadwick (2007) as activities to build the awareness, skills and knowledge base of individuals and organizations. In sustainable development, capacity-building can support change processes by integrating technological developments, investments and institutional changes to improve current and future human development needs (Bloomfield et al., 2017). Therefore, capacity-building is viewed as a component of human capital theory.

Human capital theory is the theoretical framework responsible for adapting education and development policies (Olaniyan & Okemakinde, 2008). It is embedded in the work of Smith (1776 in Mulongo, 2012) and Marshall (1890 in Mulongo, 2012). The human capital theory centres on education as an investment for the individual to acquire the necessary knowledge and skills to enable them to contribute to the development of society (Nanfosso, 2011, in Khan et al.,

2018). The investment in education contributed to enhancing the human capital stock of society. Hence, education is an investment for future returns from employment and contributes to the development of humanity and society.

Education is essential to improving resilience to climate change and attaining Sustainable Development Goals (UN, 2015; UNEP, 2013). Pressures from global warming will make it more challenging for developing countries to achieve the development targets for health and education (Brown & Westaway, 2011). Nevertheless, better-quality education would help people to develop a clear understanding of how to cope with and respond to changes in climatic conditions. Article 6 of the United Nations Framework Convention on Climate Change (UNFCCC) calls on governments to develop and implement programmes on education and awareness (United Nations, 2015).

The capacity-building activities can create opportunities for multistakeholders to engage in action-planning activities where the concerns and efforts of stakeholder groups coincide. Capacity-building is an important development strategy for enhancing the capacity of individuals and organizations. The capacity of institutions involved in climate change adaptation activities can be enhanced through (a) identifying the needs, gaps, knowledge and ways to bridge them through training; (b) identifying and sharing best practices and lessons learned; (c) giving people tools and techniques to enable themselves to solve their problems; and (d) developing experts who are knowledgeable and willing to work and share their knowledge with communities and organizations within the state (internally) and across countries (externally). There is a range of strategies for enhancing the capacity of institutions involved in the climate change adaptation activities, including identifying and sharing best practices and lessons learned, empowering people by giving them the tools and techniques to enable them to solve their problems, developing experts who are knowledgeable and willing to work to share their knowledge with communities and organizations, identifying skills that already exist, building and/or emphasizing the abilities and capacities of institutions and individual and creating platforms for exchange between stakeholders and designing "toolkits" for implementing actions. Building the mechanisms for enhancing monitoring and evaluating (M&E) the outcomes of capacity-building is a critical part of the strategy. The M&E mechanism would help organizations understand their efforts' results and engage them in adaptive management actions and programmes (Bloomfield et al., 2017).

Education and training are key capacity-building efforts to increase sustainability, institutionalizing information and knowledge about climate change adaptation. Such capacity-building efforts can be reached via context-specific or tailored education and training. There is an increased consideration of adaptation-relevant content in existing or new education and training initiatives focused on different groups and available in diverse sectors and fields. Training in the field of climate change adaptation for professionals supports better delivery of work and thus increases adaptive capacity and climate resilience. These training courses are opportunities for innovators, practitioners and entrepreneurs to find solutions to address the climate change challenges.

Furthermore, through education and training, stakeholders can be provided with climate-specific knowledge and skills for improving agriculture, fisheries, tourism and other forms of livelihood performance. The strategy can be further enhanced by (a) improving occupational safety and health, (b) the responsible use and management of pesticides and other hazardous chemicals, (c) investments in extension services, (d) provision of support for rural infrastructure and (e) developing a non-farm rural economy initiative (ILO, 2013; IPCC, IPCC, 2014a, b, c; Poschen, 2015; UN, 2015).

1.1.2 Policymaker, Leadership and Capacity Development

Policymakers have varied requirements and may require approaches designed to improve social, institutional and leadership capacity (Jayatilaka, 2003). Policymakers can apply these skills and competencies to achieve sustainable goals at local and national levels. It is hoped that improving the competencies of policymakers would provide them with opportunities for influencing decisions and discourse regarding the achievements of SDGs (Bloomfield et al., 2017). Although policymakers and their institutions acknowledge the significance of continuous follow-up support in furthering the achievements of SDGs, the demand for this support can outstretch the organization's financial resources and staffing capacity. Therefore, sustained efforts must be made by the management of the institutions to ensure an adequate, sustainable supply of financial resources and the availability of experts to implement capacity-building initiatives effectively.

1.1.3 Top-Down and Bottom-Up Approaches to Climate Change Adaptation

Adaptive capacity is the capability of actors within a structure or system to react to and initiate changes in that structure (Chapin et al., 2006; Butler et al., 2015). In capacity-building, adaptive capacity processes must be carefully designed and account for stakeholders' power dynamics to ensure their effectiveness (Ballard, 2005). There are two approaches to adaptive capacity planning the authors consider when implementing capacity-building initiatives. These approaches consist of (a) top-down and (b) bottom-up approaches. The bottom-up approach to addressing climate change vulnerability seeks to empower communities by promoting and supporting self-assessment of climate impacts through participatory rural appraisal (PRA) methods (van Aalst et al., 2008, in Butler et al., 2015). Bottom-up approaches to CCA are focused on the concept of vulnerability and therefore assume that if you address tangible vulnerability today, one unavoidably reduces future vulnerability (Burton et al., 2002). Overall, bottom-up approaches consider vulnerability as a representative of social and ecological structures that are generated by multiple factors and processes (OECD, 2013). The top-down approach involves government-level decision-making based on long-term regional climate change modelling, impact projections and adaptation strategies identified through technocratic cost-benefit analyses (Kelly & Adger, 2000). The approach has evolved through National Adaptation Programmes of Action to include national to local stakeholders (UNDP, 2010, in Butler et al., 2015). Developed nations mainly use top-down approaches to climate change adaptation because they are less prone to the impacts and are more resilient to climate change.

The top-down approach is described as a command-and-control approach. Researchers theorize that this contributed to potential maladaptation exacerbating vulnerabilities and causing negative externalities (Adger et al., 2005). The approach is criticized for considering implementation as an administrative process and usually ignores or eliminates the political aspects of the process. It is also criticized for only considering the views and decisions of the central decision-makers while neglecting the views of local stakeholders (Hjem & Hull, 1982; OECD, 2013). From a monitoring perspective, a top-down approach is defined as monitoring efforts identified within a global, international or national framework, often focusing on national and international assessments and scientific research.

The bottom-up approach is characterized as observing or monitoring efforts defined and undertaken at the local scale and brought forward to higher-level bodies, often focusing on supporting outcomes desired by a local community. Bottom-up approaches originate at the local community level, with requirements tied to management outcomes that community members or institutions seek to achieve (Danielsen et al., 2009; Commodore et al., 2017; Eicken et al., 2021). Some of the strengths of the approach include (a) being more flexible than the top-down approach in that it does not present prescriptive advice but describes the factors causing difficulties in reaching stated goals (Matland, 1995) and

(b) it ensures that strategies are flexible to enable them to adapt to local difficulties and contextual aspects of the respective programme (OECD, 2013). The bottom-up approach is faced with its share of criticisms, including policy control should be implemented by officials whose powers come from their accountability to the sovereign voters through their elected officials, i.e. in a democratic context.

Both approaches possess their distinct strengths and weaknesses. Top-down planning is sector-based and cannot easily incorporate community priorities (Sherman & Ford, 2013). Research shows that stakeholders from higher-level government and NGO institutions are more aware of long-term global drivers that communities cannot easily conceptualize and can deliver appropriate strategic responses. These stakeholders also have the mandate and resources to create and change policy that enables local-level decision-making (Adger et al., 2005, in Butler et al., 2015).

There is potential to incorporate the strengths of these approaches. Combining both approaches would highlight their strengths, minimize their weaknesses and foster greater adaptive capacity than what they may achieve when implemented in isolation from each other. Research shows that connecting top-down and bottom-up approaches would promote complementarity and fit between locally based monitoring relevant to decision-makers, community planners and large-scale observations. It will also enable the various levels of stakeholders to interact and influence project outcomes (Matland, 1995; OECD, 2013). Combining both approaches into a hybrid is more applicable when responding to policies characterized by high conflict and high ambiguity intended to highlight value or significant political shift (Suggett, 2011; OECD, 2013; Stachowiak et al., 2016; Matland, 1995). Policy implementation is most likely effective when all stakeholders at various levels can interact by providing opportunities for both central policymakers and local stakeholders on the ground to work with each other. This situation is extremely important for implementing policy and strategic programmes (OECD, 2013).

Research identifies challenges of implementing top-down and bottom-up approaches to adaptive capacity and capacity-building. It recognizes the need to promote collaborations across stakeholders and draw on diverse knowledge sources (Turnbull et al., 2013). Still, neither of the above research provides guidance on how to achieve these outcomes and/or consider how to bridge the potential power dynamics among the key stakeholders.

1.2 Leadership and Capacity-Building for Climate Change Adaptation

Actions taken today to support climate change adaptation and natural disaster resilience can have far-reaching and positive long-term effects on society and the environment (Becker & Kretsch, 2019). Inaction, on the other hand, can set a difficult and possibly irreversible course affecting numerous facets of human life now and in the future~~. Research indicates that~~ leadership is a major element in building people's resilience to mitigate and adapt to climate change (Kajsa-Stina Benulic et al., 2021). Therefore, any strategies to tackle the impact of climate change must underline the necessity for the transformations of society (European Commission, 2019), i.e. the implementation of a comprehensive and non-linear system that changes society into something qualitatively different from what it is presently (Feola, 2015; Hölscher et al., 2018; Linnér & Wibeck, 2019).

Due to the complexity, uncertainty and magnitude of the impact of climate change on society, it will present distinct challenges to people wishing to lead climate change and sustainable development-related organizations into the future (Meijerink & Stiller, 2013). Hence, it is imperative to note that climate change demands leadership throughout the sectorial and governance levels of the state (Torney, 2019; Benulic et al., 2021), leaders who would not only occupy the appropriate position within the organization but also must have the personality and character required to bring the right actors together (Becker

& Kretsch, 2019). The leader must work with stakeholders and sectors to create pre- and post-planning for enhancing dialogue among key stakeholders and sectors. Since resilience planning is a relatively emerging area in building climate change adaptive capacity, such activities allow stakeholders to understand their roles and their risks better.

~~Research has shown~~ Policymakers and practitioners are searching for solutions that do not only serve climate adaptation goals but integrate them into existing policy frameworks, a process known as mainstreaming (Uittenbroek et al., 2013). Mainstreaming is understood to stimulate the effectiveness of policymaking by combining objectives, increasing the efficient use of human and financial resources and ensuring long-term sustainable investments (Smit & Wandel, 2006). To be relevant, the current crop of climate change leaders would be required to be skilled and knowledgeable in the mainstreaming capacity-building models in policies and programmes.

1.3 Development of a Conceptual Framework for Capacity Development

A conceptual framework offers a comprehensive approach to thinking through how to engage individuals, organizations and broader systems that create incentives, i.e. the processes, resources, norms and values of institutions (William et al., 2020). The implication of new effective learning has also been stressed as necessary to build people's resilience to deal with climate change and other development calamities (Jones et al., 2018).

UNFCCC (2001) adopted a conceptual framework that lays down the basic principles, priorities, guidelines and scope for capacity-building. However, climate change has intensified significantly during the two decades following the establishment of the conceptual framework. Therefore, to tackle the complex nature of the climate challenge in a profound manner, there needs to be a persistent injection in human and financial resources concerning the capacity-building needs of key stakeholders and their institutions (Khan et al., 2018).

The authors reviewed various conceptual frameworks on capacity-building for climate change adaptation to determine their suitability and relevance to building the capacity of institutions and across sectors. They realized that none of the frameworks reviewed was appropriate for the task. This begs the need to create a new conceptual framework that is more appropriate and relevant to the capacity-building needs of individuals and their institutions in developing countries. The new conceptual framework is called "Climate Change Adaptation Framework for Capacity-Building in Developing Countries". It consists of four steps, each comprising a set of strategic tools to operationalize the respective stage of the framework.

1.4 The Purpose of the Book: Aim and Objective

1.4.1 Aim of the Book

Through the utilization of case studies, the authors collect data on the relevance and effectiveness of the strategic actions implemented by NGOs, IGOs and governments at national, regional and international levels to enhance the capacity and governance of these institutions to deliver climate change adaptation and sustainable development programmes in developing countries. The book provides insights to policymakers, community leaders, students and researchers working on climate change adaptation and resilience-building practices and strategies in vulnerable communities, including small island developing states and least developing states, focusing on innovative practices to build their institutional capacity utilizing participatory bottom-up and top-down approaches.

The book's overall purpose is to provide insights and to learn about the unique contributions of institutions to building people's capacities to adapt to the impact of climate change and other development calamities in developing countries.

1.4.2 Objectives of the Book

- Present case studies on actions, approaches and practices implemented by organizations to enhance the capacity of the institutions involved in climate change adaptation and other development programmes, including top-down and bottom-up approaches.
- Review conceptual frameworks, models and theories on capacity-building for climate change adaptation and develop an applicable and relevant capacity-building conceptual framework to build the resilience of people at local, regional and international levels.
- Identify the strengths, weaknesses, effectiveness, lessons learned and best practices from the past capacity-building actions-oriented case studies implemented by institutions as a strategy for learning how to utilize these practices and models to enhance capacity-building initiatives to address future capacity-building needs in developing countries.
- Make recommendations to enhance the institutional capacities of organizations to effectively build the resilience of people and their communities to respond to climate change in developing countries.

1.5 Contribution of the Book to New Knowledge

The cases presented in this book underlie the need for adequate knowledge and skills to prepare policymakers and practitioners to address their vulnerability, resilience and adaptation to climate change from a top-down and bottom-up perspective/approach. The book offers a set of strategies for enhancing the knowledge and skills of policymakers, students and practitioners on actions, approaches and practices to be implemented by organizations to enhance the capacity of the institutions involved in climate change adaptation and other development programmes, including top-down and bottom-up approaches, in SIDS and other developing countries in Africa, Caribbean and Asia.

Consequently, developing this framework will also contribute substantially to strengthening the capacity and capability of institutions in SIDS and other developing countries to build the resilience of these institutions to deal with the impact of climate change. The research is situated within the framework of sustainable capital theory focusing specifically on human capital theory (ILO, 2010; UNDP, 2015). It is framed within the context of enhancing human capital through skills, knowledge and livelihood needs to build resilience to adapt to climate change.

References

Adger, N. W., Arnell, N. W., & Tompkins, E. L. (2005). Successful adaptation to climate change across scales. *Global Environmental Change, 15*(2), 77–86. Available at: http://www.sciencedirect.com/science/article/pii/S0959378004000901. Accessed 11 July 2014.

Ballard, D. (2005) Using learning processes to promote change for sustainable development. *Action Research*. 3(2), 135–156.

BBC. (2021a). *Climate change: IPCC report is 'code red for humanity'*. By Matt McGrath Environment correspondent.

BBC. (2021b). One in three trees face extinction in the wild, says new report; BBC News. *Helen Briggs, Science and environment Correspondence, September, 1*, 2021.

Becker, A., & Kretsch, E. (2019). The leadership void for climate adaptation planning: Case study of the port of providence (Rhode Island, United States). *Frontiers of Earth Science, 7*, 29. https://doi.org/10.3389/feart.2019.00029

Benulic, K.-S., Kropf, M., Linnér, B.-O., & Wibeck, V. (2021). The meaning of leadership in polycentric climate action. *Environmental Politics*. https://doi.org/10.1080/09644016.2021.1970087

Bloomfield, G., et al., (2018) Capacity building to advance the United Nations sustainable development goals: An overview of tools and approaches related to sustainable land management. *Journal of Sustainable Forestry. 37*(2):157-177.

Brown, K., & Westaway, E. (2011). Agency, Capacity and Resilience to Environmental Change: Lessons from Human Development, Well-Being and Disaster. *Annual Review of Environment and Resources., 36*, 321–342.

Burton, I., Huq, S., Lim, B., Pilifosova, O. & Schipper, E.L. (2002) From impacts assessment to adaptation priorities: the shaping of adaptation policy. *Climate Policy, 2*(2–3), 145–159.

Butler, J., Wise, R., Skewes, T., Bohensky, E., Peterson, N., Suadnya, W., Yanuartati, Y., Handayani, T., Habibi, P., Puspadi, K., Bou, N., Vaghelo, D., & Rochester, W. (2015). Integrating top-down and bottom-up adaptation planning to build adaptive capacity: A structured learning approach. *Coastal Management, 43*, 346–364. Taylor & Francis Group, LLC; ISSN: 0892-0753 print /1521-0421 online. https://doi.org/10.1080/08920753.2015.1046802

Chapin, F. S., Lovecraft, A. L., Zavaleta, E. S., Nelson, J., Robards, M. D., Kofinas, G. P., Trainor, S. F., Peterson, G. D., Huntingdon, H. P., & Naylor, R. L. (2006). Policy strategies to address sustainability of Alaskan boreal forests in response to a directionally changing climate. *Proceedings of the National Academy of Sciences of the United States of America, 103*, 16637–16643.

Clemencon, R. (2016). The two sides of the Paris Climate Agreement: Dismal failure or historic breakthrough? *Journal of Environment and Development, 25*(1), 3–34. SAGE.

Commodore, A., Wilson, S., Muhammad, O., Svendsen, E., & Pearce, J. (2017). Community-based participatory research for the study of air pollution: A review of motivations, approaches, and outcomes. *Environmental Monitoring and Assessment, 189*, art. 378.

Danielsen, F., et al. (2009). Local participation in natural resource monitoring: A characterisation of approaches. *Conservation Biology, 23*, 31–42.

Eicken, H., Danielsen, F., Sam, J. M., Fidel, M., Johnson, N., Poulsen, M. K., Lee, O. A., Spellman, K. V., Iversen, L., Pulsifer, P., & Enghoff, M. (2021). Connecting top-down and bottom-up approaches in environmental observing, *BioScience, 71*(5), 467–483.

European Commission. (2019). Communication from the Commission to the European Parliament, the European Council, the Council, The European Economic and Social Committee and the Committee of the Regions. The European Green Deal, COM. (2019)*640*, final.

Feola, G. (2015). Societal transformation in response to global environmental change: A review of emerging concepts. *Ambio, 44*(5), 376–390. https://doi.org/10.1007/s13280-014-0582-z

Gordon, J., & Chadwick, K. D. (2007). *Impact assessment of capacity building and training: assessment framework and two case studies*.

Hjem, H., & Hull, C. (1982). Implementation research as empirical constitutionalism. *European Journal of Political Research., 10*, 105–116.

Hölscher, J. K., Wittmayer, J. M., & Loorbach, D. (2018). Transition versus transformation: What's the difference? *Environmental Innovation and Societal Transitions, 27*, 1–3. https://doi.org/10.1016/j.eist.2017.10.007

ILO. (2010). *A skilled workforce for sustainable growth: A G20 training strategy*. International Labour Office.

ILO. (2013). *Sustainable development, decent work and green jobs*. In *International Labour Conference, 102nd Session*. International Labour Office.

IPCC. (2014a). Synthesis report. In *Contribution of working groups I, II and III to the fifth assessment report of the Intergovernmental Panel on Climate Change*, IPCC Geneva, Switzerland, p. 151.

IPCC. (2014b). Summary for policymakers: Climate change impacts, adaptation and vulnerability. In C. B. Barros, V. R. Dokken, D. Mach, K. J. Mastrandrea, M. D. Bilir, T. E. Chatterjee, M. Ebi, & K. L. Estrada (Eds.), *Contributions of the working group ii to the fifth assessment report, Field*. IPCC. https://www.ipcc.ch/assessment-report/ar6/

IPCC. (2014c). Climate change: Synthesis Report. In *Contribution of working group 1, 11, 111 to the fifth assessment report of the Intergovernmental Panel on Climate Change*, IPCC Geneva, Switzerland, p. 151.

IPCC. (2022). Climate change 2022 mitigation of climate change, summary for policy makers: Working group 111 contribution to the sixth assessment report of the International Panel on Climate Change (AR6 WG 111), IPCC.

Jayatilaka, W. (2003). Capacity-building for sustainable development. *Journal of the National Science Foundation of Sri Lanka, 31*, 81–97.

Jones, L., Harvey, B., Cochrane, L., Cantin, B., Conway, D., Cornforth, R. J., et al. (2018). Designing the next generation of climate adaptation research for development. *Regional Environmental Change, 18*, 297–304.

Kelly, P. M., & Adger, W. N. (2000). Theory and practice in assessing vulnerability to climate change and facilitating adaptation. *Climatic Change, 47*, 325–352.

Khan, M. R., Timmons Roberts, J., Huq, S., & Hoffmeister, V. (2018). *The Paris framework for climate change capacity-building*. Routledge/Taylor and Francis Group/Earthscan.

Linnér, B.-O., & Wibeck, V. (2019). *Sustainability transformations: drivers and agents across societies*. Cambridge University Press.

Matland, R. (1995). Synthesising the implementation literature: The ambiguity-conflict model of policy implementation. *Journal of Public Administration Research and Theory, 5*(2), 145–174.

Meijerink, S., & Stiller, S. (2013). What kind of leadership do we need for climate change adaptation? A Framework for analysing leadership objectives, functions and tasks in climate change adaptation. *Environ Plan C, 31*, 240–256. https://doi.org/10.1068/c11129

Mulongo, G. (2012). *Human Capital in Education: Principles, critiques and current thinking*. Institute of Education, University of London.

Nurse, L. A., McLean, R. F., Agard, J., Briguglio, L. P., Duvat-Magnan, V., Pelesikoti, N., Tompkins, E., Webb, A. (2014). Small islands. In Barros, V.R., Field, C.B., Dokken, D.J., Mastrandrea, M.D., Mach, K.J. Bilir, T.E., Chatterjee, M., Ebi, K.L., Estrada, Y.O., Genova, R.C., Girma, B., Kissel, E.S., Levy, A.N., MacCracken, S., Mastrandrea, P.R., White, L.L. (Eds.), *Climate change 2014: Impacts, adaptation, and vulnerability*. Part B: Regional aspects. Contribution of working group II to the fifth assessment report of the Intergovernmental Panel on Climate

Change, Cambridge University Press, Cambridge, UK and New York, pp. 1613–1654.

OECD. (2013). *The nature of policy change and implementation: A review of different theoretical approaches. Organisation for Economic Cooperation and Development.*

Olaniyan, D. A., & Okemakinde, T. (2008). Human capital theory: Implementation for education and development. *Pakistan Journal of Social Development, 5*(5), 479–483.

Poschen, P. (2015). *Decent work, green jobs and the sustainable economy: Solution for climate change development.* International Labour Organisation.

Sherman, M., & Ford, J. (2013). Stakeholder engagement in adaptation interventions: An evaluation of projects in developing nations. *Climate Policy*. http://doi.org/r45. Accessed 15 Dec 2014.

Smit, B., & Wandel, J. (2006). Adaptation, adaptive capacity, and vulnerability. *Global Environmental Change, 16*, 282–292. https://doi.org/10.1016/j.gloenvcha.2006.03.008

Stachowiak, S., Robles, L., Habtermanion, E., & Maltry, M. (2016). *Beyond win: Pathway for policy implementation.* ORS Impact/The Atlas Learning Project.

Suggett, D. (2011). *The implementation challenge: Strategy is only as good as the execution* (State Services Authority Occupational Paper No. 15).

Torney, D. (2019). Follow the leader? Conceptualising the relationship between leaders and followers in polycentric climate governance. *Environmental Politics, 28*(1), 167–186.

Turnbull, M., Sterrett, C. L., & Hilleboe, A. (2013). *Toward resilience: A guide to disaster risk reduction and climate change adaptation.* Practical Action Publishing Ltd.

Uittenbroek, C. J., Janssen-Jansen, L. B., & Runhaar, H. A. (2013). Mainstreaming climate adaptation into urban planning: Overcoming barriers, seizing opportunities and evaluating the results in two Dutch case studies. *Regional Environmental Change, 13*, 399–411. https://doi.org/10.1007/s10113-015-0886-y

UNFCCC (2001). *Report of the Conference of the Parties on its seventh session, held at Marrakesh from 29 October to 10 November 2001.* Addendum. Part two: Action taken by the Conference of the Parties. Volume I. Available at: https://unfccc.int/sites/default/files/resource/docs/cop7/13a01.pdf

UNDP. (2015). *Human development report.* Work for Human Development, United National Development Programme (UNDP).

United Nations (UN). (2015). Resolution adopted by the General Assembly on September 25 2015. /RES/70/1 Transforming our world: The 2030 Agenda for Sustainable Development.

United Nations (UN). (2021). Secretary-general's statement on the IPCC working group 1 report on the physical science basis of the sixth assessment; United Nations New York; Statements on 09 August 2021

United Nations Environment Programmes. (2013). *Identifying emerging issues from the perspective of Small Island Developing States (SIDS).* In *Expert group meeting co-hosted by United Nations Environment Program (UNEP) and the United Nations Department of Economic and Social Affairs (DESA), 14–16 May 2013.* United Kingdom.

United Nations Framework Convention on Climate Change. (2021). Implementation of the framework for capacity-building in developing countries Synthesis report by the secretariat: FCCC/SBI/2021/3: Distr.: General 10 March 2021.

Van Aalst, M. K., Cannon, T. & Burton. I. (2008). Community level adaptation to climate change: the potential role of participatory community risk assessment. *Global Environmental Change, 18*, 165–179

2 Literature Review

Abstract

The chapter analyses the current research on building capacity around climate change and its impact on achieving Sustainable Development Goals (SDGs). The chapter explores the literature on climate change, specifically highlighting the impact of climate change on developing countries' roadmap to achieving the UN SDGs. It investigates strategies' definition, effectiveness and relevance to address development challenges in LDCs and developing countries. The literature examined the gaps, challenges and successes in the journey from the UNFCCC COP21 Paris Climate Agreement to the Edinburgh, Scotland, UK, COP26 in considering keeping global warming within 1.5 and 2 °C and the impact of attaining SGDs. The chapter further explores top-down and bottom-up approaches, sustained leadership and participatory approaches/models utilized by agencies to enhance institutional capacity to build their resilience to adapt to climate change and other development challenges.

The chapter also reviews existing knowledge to underpin the relevance and appropriateness of the current conceptual framework for building the capacity of institutions to adapt to climate change. The review shows that the conceptual framework designed by some schools of thought needs to be more appropriate and relevant to building the capacity of individuals and institutions in developing countries. In this chapter, a new conceptual framework is presented by utilizing integrated top-down and bottom-up participatory integrated approaches to build the capacity of institutions to adapt to the impact of climate change in developing countries. Gender factors and consequences are considered in the development of the new framework.

Keywords

Conceptual framework · Least developing country · Sustainable Development Goals · Climate change adaptation · Top-down approach · Bottom-up approach · Participatory approach · Human capacity · Gender · Leadership

This chapter analyses the current research on building capacity around climate change and its impact on achieving Sustainable Development Goals (SDG). The chapter explores the literature on climate change, specifically highlighting the impact of climate change on developing countries' roadmaps to achieving the UN SDGs. It investigates strategies; definition, effectiveness and relevance to address development challenges in LDCs and SIDS. The literature examined the

A. Simmons, A.-A. Simmons, *Building Institutions for Resilience*, Sustainable Development Goals Series, https://doi.org/10.1007/978-3-031-28881-4_2

gaps, challenges and successes in the journey from the UNFCCC COP21 Paris Climate Agreement to the Edinburgh, Scotland, UK, COP26 in considering keeping global warming within 1.5 and 2 °C and the impact of attaining the SGDs. The chapter further explores top-down and bottom-up approaches, sustained leadership and participatory approaches/models utilized by agencies to enhance institutional capacity and build their resilience to adapt to climate change and other development challenges.

The chapter also reviews existing knowledge to underpin the relevance and appropriateness of the existing conceptual framework for building the capacity of institutions to adapt to climate change. The review shows that the conceptual framework designed by some schools of thought is not appropriate and relevant to building the capacity of individuals and institutions in developing countries. In this chapter, a new conceptual framework is presented by utilizing integrated top-down and bottom-up participatory approaches to build the capacity of institutions to adapt to the impact of climate change in developing countries. Gender factors and consequences are considered in the development of the new framework.

The persistent progression of climate change (IPCC, 2014a, b, c, 2022) necessitates innovations in governance systems to deal effectively with the impact of climate change. Therefore, enhancing the institutional capacity of organizations/institutions to deliver climate change adaptation actions is an important indicator of successful climate adaptation and mitigation programmes. Stakeholders must ensure that efforts are made to enhance public and private institutions' institutional, governance and accountability mechanisms to effectively build community and national resilience to adapt to climate change's impact. The literature shows that such a decision-making process should be instituted by a configuration of actors, including state, private and civil society/non-governmental organizations. The appropriate institutional arrangements to achieve desirable outcomes must be vigorously instituted (Adger, 2003; Ojwang et al., 2017). Such arrangements possess the substantial potential to decrease the impacts of climate change at all levels (Pahl-Wostl et al., 2013). Implementors of innovations in multilevel governance systems ought to be cognizant of the political realities of LDCs and SIDS to identify context-sensitive, bottom-up and community-led solutions to produce equitable, efficient and effective outcomes (Adger, 2003; UNDP, 2009).

Vulnerability is a critical part of the impact of climate change (IPCC, 2022). Therefore, research on vulnerability and strategies for building resilience needs to carefully explore and augment the role of governance (Williams et al., 2020). The effectiveness of multilevel governance understood as the sharing of power across global institutions, regional organizations and national and sub-national governance (Schakel et al., 2015) is often impeded in climate policy processes through a lack of coordination between national and local government levels. Importantly, national policy planners would insufficiently consider local capacities for implementing climate change adaptation measures. Therefore, it is argued that effective multilevel governance for climate change adaptation necessitates close collaboration and cooperation between these political and organizational structural dimensions (Leck & Simon, 2013).

As the impacts of climate change are primarily experienced locally, locally driven action is important (IPCC, 2022). "The focus on local governance, understood as the political and institutional processes through which decisions are taken and implemented in a specific sub-national geographic region, is reflected in the National Adaptation Plans (NAPs) adopted under the United Nations Framework Convention on Climate Change (UNFCCC)" (Etongo & Gill, 2022, pg 2), in which significant consideration of local stakeholders to implement climate change adaptation is strongly recommended (Di Gregorio et al., 2019; UNFCCC, 2017).

In accepting responsibility for building the local capacity to adapt to climate change, local governance is tasked with, among other issues, incorporating climate risks in the development planning process, mobilizing resources for adaptation, adjusting building and land use regulations to consider climate risks and enhancing disaster risk preparedness and management, response and recovery (Corfee-Morlot et al., 2011; Reisinger et al., 2011).

(a) *Building Institutional Capacity for Climate Change Adaptation*

The models most cited in the literature on capacity-building were (a) the Ecological Model and (b) Interactive Systems Framework for Dissemination. The most frequently cited framework was Bloom's Taxonomy of Learning (Bergeron et al., 2017). The literature on capacity-building approaches identified included training, technical assistance, knowledge networks and professional coaching (Bergeron et al., 2017).

Institutional capacity is important for successfully implementing climate change adaptation actions (Dany et al., 2015). Institutional capacity development is a severe deficiency in several LDCs, SIDS and other developing countries with limited financial resources and viable human capital (Callaghan, 2015). Therefore, any effort to increase capacity should be based on understanding the existing capacity levels within the institution or at the state level. The future priorities for tackling capacity-building can then be subsequently identified at the next stage.

Dany et al. (2015) identify four capacity development elements regarding climate change adaptation. These include (1) financial resources, (2) cooperation and coordination of stakeholders, (3) availability and quality of information on vulnerability and adaptation to climate change and (4) the level of understanding of climate change vulnerability and adaptation. As highlighted above, research showed that informants perceived the capacity for climate change adaptation in developing countries to be very constrained, especially regarding financial resources and cooperation. These critical factors were ranked the highest climate change capacity priority for addressing the current climate change challenge in developing countries (Dany et al., 2015).

Dany et al. (2015) stated that institutional capacity constraints were generally related to weak governance within governments and development organizations to implement climate change adaptation initiatives. Due to the critical role that higher education institutions can play in achieving climate change adaptation, the research findings state that the absence of local higher education institutions in climate change adaptation activities should be addressed. Therefore, the participation of such educational institutions would provide an important mechanism to advance both capacity development and partnerships and coordination among and between organizations and relevant sectors (Dany et al., 2015).

(b) *Climate Change Adaptation and Capacity-Building*

Mainstreaming climate change adaptation into policy development is an essential strategy to implement across all sectors of society (Chevallier, 2010; England et al., 2018; Mataya et al., 2020; IPCC, 2022). There is a clear understanding by climate change practitioners and policymakers that adaptation is not a one-off action but a process that requires adaptive management reflecting unfolding climate impacts, the prescriptive nature of risk tolerance and the tipping points between both phenomena (Wise et al., 2014). Although knowledge of adaptation is necessary for the successful implementation of climate change adaptation pre-programmes, there is the absence of an exact blueprint on what should constitute climate change adaptation actions for an area or region due to the context-specific and process-based nature of implementing adaptation action (Williams et al., 2015; Mataya et al., 2020).

Although reiterative reflections within adaptation processes help to highlight the most appropriate adaptive pathways (Tschakert & Dietrich, 2010), on a broader scale, social learning can include reflecting, knowledge sharing and providing opportunities for the co-creation of knowledge so that change goes beyond individuals into the community- and national-level practices and applications (Ensor & Harvey, 2015; Mataya et al., 2020). Meaningful learning that evolves from earlier applications of climate change adaptation practices is that adaptation knowledge needs to be shared widely to build adaptation capacity (Mataya et al., 2020).

(c) *Capacity-Building and the Sustainable Development Goals*

International leaders met at the United Nations Headquarters in New York in September 2015 to announce an ambitious and comprehensive agenda of 17 goals and 169 targets to achieve transformational social, economic and environmental change by 2030. These goals are called Sustainable Development Goals (SDGs) and provide a comprehensive framework for addressing the development needs on a global, regional and national scale (UNDP, 2015). Capacity-building is one of the targets in goal 17 of the Sustainable Development Goals, strengthening the implementation and revitalization of global partnerships for sustainable development (https://sdgs.un.org/goals/goal17).

Article 6 of the United Nations Framework Convention on Climate Change (UNFCCC) acknowledges the value of learning and capacity-building to react to the global climate change challenge. UNFCCC pursues this by accelerating the training of "scientific, technical, and managerial personnel, as one way of implementing sustainable climate change adaptation programmes" (UN/UNFCCC/Article 6, 1992; Mataya et al., 2020). The UNFCCC, in keeping with its philosophy of collective but differentiated responsibilities, makes pledges to assist developing countries with the transfer of skills, knowledge and technologies to enable sustainable ecological adaptation.

Capacity-building is critical for achieving Sustainable Development Goals in all developing countries, specifically in SIDS. Capacity development and building are used widely and interchangeably across development, health, agricultural and environmental sectors as vital strategies for achieving development programme goals (Jayatilaka, 2003; OECD, 2006; Bloomfield et al., 2017). The relevance of capacity-building to the development of developing countries and SIDS is highlighted in the SDG target 17.9 to "Enhance international support for implementing effective and targeted capacity-building in developing countries to support national plans to implement all the sustainable development goals, including through North-South, South-South, and triangular cooperation" (https://sdgs.un.org/topics/capacity-development).

2.1 Defining Capacity-Building

Institutional strengthening, capacity-building and organizational development have been focal points of development practice since the 1950s (Mackenzie & Gordon, 2016, in Shakya et al., 2018). Most developing countries have recently indicated capacity-building as a condition for implementing their intended nationally determined contributions (INDCs) under the Paris Agreement (Khan et al., 2019). Although the definitions of capacity-building may vary theoretically, there is a common intention for implementing capacity development activities, i.e. to strengthen an organization's effectiveness in achieving its goals (Whittle et al., 2012). The work on capacity development takes on a range of disciplines, including governance, leadership development and institutional and behavioural economics.

Capacity-building is a concept that is broader than education and training and institutional development since it emphasizes the environment and context within which individuals, organizations, institutions and societies operate and interact (Enemark, 2003; Khan et al., 2019). It is suggested that capacity development occurs at three levels: individual, organizational and societal (OECD DAC, 2019; CHF, 2008; UNDP, 2009; World Bank, 2009). However, some sources have differentiated between organizational and institutional capacities (North, 1990), while others add capacity development at the global level (Pearson, 2011a). The main element to success in all these levels is ownership of the process and products (Bolger, 2000; Lafontaine, 2000; Lopes & Theisohn, 2003; Khan et al., 2019). The UN Commission for Africa emphasizes that developing countries must be the ones to lead the initiatives on capacity-building (Khan et al., 2019).

Capacity-building is defined as endeavours and actions to build individuals' and organizations' understanding, skills and knowledge base to achieve their development goals (Bloomfield et al., 2017). It is also described by Gordon and Chadwick (2007) as activities to build the awareness, skills and knowledge base of individuals and organizations. In sustainable development, capacity-building can support change processes by integrating technological developments, investments and institutional changes, to enhance and expand current and future human development needs (Bloomfield et al., 2017).

There is a tendency for most development agencies to use the concept of capacity-building and capacity development interchangeably. The word "development" acknowledges existing capacities and emphasizes strengthening what is already there rather than starting something new. Some researchers argued that there is no basic difference between these two terms (Khan et al., 2019), while others opined that capacity-building is regarded as starting from nought, while capacity development is understood as having a base from which to start the process (Kuhl, 2009; Pearson, 2011a). Freeman (2010, p. 17) supports this view and argues that "capacity-building sits with the more traditional top-down knowledge transfer and technical cooperation approaches".

From an international development context, numerous dialogues viewed capacity-building as the provision of training. It is also classified as a necessary phenomenon that enhances further than training, i.e. knowledge transfer that leads to better delivery or implementation or ability to address a specific problem or challenge (Virji et al., 2012; Mataya et al., 2020). Therefore, capacity-building emphasizes improving knowledge, logic and practical skills through training and incentivizing institutions (Boyd et al., 2014). Most developing countries cannot deal with the impact of climate change; therefore, enhanced efforts in capacity-building are necessary for developing countries and SIDS to build the resilience of institutions to adapt to the impact of climate change and the delivery of the UN SDGs.

The authors shared the definitions of Mataya et al. (2020) on capacity-building as enhancing skills and ability that translates into better adaptation. This definition of capacity-building recognizes that applying new knowledge and skills relies upon the institutional environment in which individuals attempt to utilize such capacity (Mataya et al., 2020). Other definitions of capacity development include (Table 2.1):

Table 2.1 List of definitions of capacity-building

UNDP	Capacity development: The process through which individuals, organizations, institutions and societies develop abilities to perform functions, solve problems and set and achieve objectives
OECD	Capacity development: The process by which individuals, groups and organizations, institutions and countries develop, enhance and organize their systems, resources and knowledge, all reflected in their abilities, individually and collectively, to perform functions, solve problems and achieve objectives (OECD, 2006)
CIDA	Capacity-building: A process by which individuals, groups, institutions, organizations and societies enhance their abilities to identify and meet development challenges in a sustainable manner (CIDA, 1996, in Lusthaus et al., 1999)
USAID	Capacity-building: An ongoing evidence-driven process to improve the ability of an individual, team, organization, network, sector or community to create measurable and sustainable results in terms of effectively "apply[ing] its skills, assets and resources to achieve its goals" (USAID, 2012)

2.2 Approaches to Capacity-Building

Capacity-building invokes a meaning of something practical in nature. According to the World Bank (2005) report, capacity comes with no accepted or tested body of theory. It is imperative to view approaches to building capacity from both theoretical and practical lenses. From a theoretical perspective, capacity-building is viewed as the development of capabilities standpoint (Sen, 1997, 1999; Khan et al., 2019). Sen has recognized a connection between the theories of capacity and human capital. The approach is valuable in hypothesizing the process and practice of capacity-building. Sen's concept of capabilities can be viewed from individual and group perspectives as the freedom to choose from alternate societal choices. Therefore, for an individual to acquire such capabilities, education and training provide the avenue for achieving such benefits.

The human capital theory centres on education as an investment for the individual to acquire the necessary knowledge and skills to enable him/her to contribute to the overall development of society (Nanfosso, 2011, in Khan et al., 2019). Therefore, education investment contributed to enhancing society's human capital stock. Overall, an investment in education is viewed as an investment for future returns from employment and contributing to the development of humanity and society.

According to Mataya et al. (2020), capacity development broadly comprises three stages: (a) needs assessment, (b) strategies and actions and (c) monitoring and evaluation. The needs assessment stage may comprise stock-taking of the existing skills and capabilities at the organizational and individual levels, followed by a comparison with identified current and future skills and competency needs of the organization.

The strategies and actions stages relate to the description of the requirement of a training plan at organizational and individual levels to address the skills and competency gaps in the organization. These strategies would incorporate various approaches to building skills and cognitive capacity at varying organizational levels, aiming to enhance awareness and increase knowledge and skills development for practice (Harrison, 2009, in Mataya et al., 2020).

There are different approaches to capacity development designed to focus on the specific needs of a particular stakeholder group, while other approaches may alternatively include tactics intentionally bringing together different stakeholder types to facilitate exchanges while addressing multiple needs of the stakeholders and the institutions (Rao et al., 2014; Bloomfield et al., 2017). The capacity-building activities can create opportunities for multistakeholder to engage in action-planning activities where the concerns and efforts of all three of these stakeholder groups coincide.

Capacity-building programmes can be formal and/or non-formal. Modern capacity-building activities are conducted in the blended field and online courses. Some specialized courses are short-duration and either non-degree or for college credit, aiming to provide practical training based on e-learning content backed with fieldwork. These courses are designed for students to develop individual projects to understand ecological restoration throughout multiple modules. Scientific and technical issues are addressed by a group of guest instructors and project advisers (Bloomfield et al., 2017). Online content of capacity-building courses may include theory-based written material, short videos on techniques and approaches, pre-recorded presentations by international experts, a virtual library and a series of individual exercises. Several advisors may be available to assess student progress and suggest approaches, readings, contacts and complementary activities to improve individual projects during each course. Courses include fieldwork at selected demonstration sites where students use tools and techniques for site diagnosis, ecological restoration planning and social and community assessment (Bloomfield et al., 2017).

Capacity-building courses can also form formal education, such as accredited degrees and diplomas from a recognized tertiary institution. There are diverse training approaches, from workshops to work shadowing to mentoring and coaching (Mataya et al., 2020). Based on empirical evidence from multiple locations, concern has been communicated over the ability of short-term training to effectively build capacity for adaptation (Mataya et al., 2020). The book explores a set of case studies on bottom-up and top-down approaches to capacity-building, including the JEMS Progressive Community Organization (bottom-up), the Government Community Development Division Programmes (top-down), St. Vincent and the Grenadines, the Government of Sierra Leone/UNDP Youth Employment and Empowerment Project (Hybrid Approach and the

Commonwealth Youth Programme (Hybrid Approach)) and others to add value to the debate.

2.3 Strategies for Building Capacity Development

Capacity development is considered an essential educational approach for improving the capacity of individuals and organizations to build resilience to adapt to climate change. Strategies for enhancing the capacity of institutions include:

- Identifying needs, gaps and knowledge through stakeholder engagement and ways to bridge them through training.
- Categorizing stakeholder needs and gaps, sharing best practices and lessons learned.
- Recognizing the skills that already exist and other building or emphasizing these abilities and individual capacities.
- Equipping people with appropriate tools and techniques to enable them to solve their problems.
- Developing experts who are knowledgeable and willing to work and share their knowledge with communities and organizations.
- Creating platforms for exchange between stakeholders.
- Designing "toolkits" for implementing actions. Most importantly, building the mechanisms for enhancing monitoring and evaluating capacity-building outcomes. This mechanism can help organizations understand the results of their efforts and engage them in adaptive management actions and programmes (Bloomfield et al., 2017).

2.4 Strengths and Weaknesses of Capacity-Building Methodologies

Training workshops, mentorship, networking, self-driven learning and publication are some of the most frequently used strategies and methodologies associated with capacity-building. There is a range of activities that support capacity-building. These include targeted events, debates, sharing information through web platforms and portals, newsletters, reports, policy briefs, videos, brochures, projects, etc. There are also different types of capacity-building, which can be categorized as follows:

- Education (e.g. through schools, universities and other education service providers).
- Training (e.g. courses, seminars, webinars, e-learning).
- Networking (e.g. conferences, workshops, sharing platforms, communities of practice, networks of excellence).
- Targeted coaching exercises.
- Technical assistance (e.g. expert missions, twinning).
- The attention of groups at risk.

These methodologies have their strengths and limitations. It is essential to note that partnerships and sharing experiences and knowledge contribute to further increasing capacity-building and learning over time (Mataya et al., 2020).

Education and training are critical capacity-building efforts to increase sustainability and institutionalize information and knowledge about climate change adaptation (Mataya et al., 2020). Such capacity-building efforts can be reached via context-specific or tailored education and training programmes. There is an increased consideration of adaptation-relevant content in existing or new education and training initiatives focused on different groups and available in diverse sectors and fields. There have been increases in education efforts at schools and universities and specific programmes, courses and events. Training in climate change adaptation for professionals supports better delivery of work and thus increases adaptive capacity and climate resilience of institutions at community and national levels.

2.4.1 Training Workshops

Workshops are deemed a standard method of short-term training, perceived by stakeholders to build practitioners' capacity to enhance skills in specific areas of interest directly linked to adaptation practice. The design of workshops can encompass many different types of training approaches, tools and methods. Research shows that participants, supervisors and trainers prefer a combination of training approaches, such as presentations, discussions, field practice, field visits, documentaries on best practices and expert talks emphasizing tailored examples. Trainers perceived sessions involving a combination of training methodologies to boost cognitive and practical skills. Training workshops that are practical in their orientation and delivery allow participants to give positive feedback regarding their experience in the training exercises. Various practical participatory methods were valuable in capacity-building training workshops (Mataya et al., 2020).

2.4.2 Targeted Coaching and Mentorship

A fundamental capacity-building approach is mentorships for skills enhancement, awareness-raising and information sharing. When used effectively, mentorship can play a significant role in enhancing the career prospects of practitioners because it enables the person to keep in touch with experts in the field and gain valuable understanding and knowledge on climate change adaptation. However, the data shows that mentorship is an infrequently used training method in developing countries (Mataya et al., 2020). Although trainers endorsed the added value of mentorship as a capacity-building methodology, it is challenging to implement because of financial constraints. These constraints are (a) trainers are engaged on a consultancy basis with a fixed-term assignment and are not available to mentor participants after the assignment is completed and (b) donors who fund training mainly contribute funding for the actual delivery of the training and not for pre-training activities such as training needs assessment and/or post-training follow-ups such as mentoring of participants to support the development and implement action plans (Mataya et al., 2020).

2.4.3 Networking

Networking was also viewed as a vital adaptation process. Participants refer to contacts for networking purposes as significant because it helps them to keep abreast of information on recent publications and emerging issues such as climate change. Although networking is viewed as an important capacity-building tool, research shows that only a minority of the trainers maintain contact with participants by following up on the action plans developed after the training workshops. It is critical to note that although many trainers expressed their interest in and support for the development of networking and the provision of ongoing support, the stakeholders highlighted the challenges of money and time as constraints for the effective implementation of this approach. However, it is important to note that such challenges can be overcome using "free-to-use" online platforms and social media where trainers and learners can continue interacting and sharing ideas. However, there are limitations to using online spaces to foster communities of practice. This methodology can be ineffective where there is absence of a champion in the community to drive the forum. A trainee participant should drive the forum (Mataya et al., 2020).

2.4.4 Self-Driven Learning and Publication

Researchers mention necessary learning tools such as on-the-job training, research, reading and internet-based sources. These approaches remain little pursued by practitioners and policymakers alike in developing countries. The low levels of stakeholders' engagement with self-driven learning reflect a lack of resources, with most government officials having poor or no internet access. It could also reflect a lack of awareness of the wide range of web-based learning tools that are available for capacity-building regarding climate change adaptation (Mataya et al., 2020).

2.5 Capacity Development and Policymakers

Policymakers have diverse needs and require approaches to review and accelerate social, institutional or leadership capacity (Jayatilaka, 2003). Policymakers can apply these skills and competencies to achieve sustainable goals at local and national levels. It is also hoped that increases in the competencies of policymakers would provide them with opportunities for influencing decisions and discourse regarding the achievements of SDGs (Bloomfield et al., 2017). Although policymakers and their institutions acknowledge the significance of continuous follow-up support in furthering SDG achievements, the demand for this support can outstretch the organization's financial resources and staffing capacity. Another important question to consider, especially in developing countries' capacity-building process, is to what extent does political interference influence the implementation of policies?

2.6 Monitoring and Evaluation of Capacity-Building for CCA

Monitoring and evaluation are relevant aspects of the capacity-building programmes within an institution. It is important for them to ensure that they are achieving their programme goals and objectives and allowing for the implementation of adaptive management and adjusting programme strategies within the organization over time (Connolly & York, 2002). Research shows that evaluating the effectiveness of such capacity-building programmes can be impeded by classical monitoring. Despite suggestions for evaluating improvement in the different learning, skills and attitudes of beneficiaries of such programmes, evaluation tools have been inadequate (Shaha et al., 2004, in Mataya et al., 2020).

2.7 Benefits of Capacity-Building for Climate Change Adaptation

The stakeholders and researchers recognize education's role in providing context and foundational knowledge. They believe that individuals educated to the tertiary level have a greater ability to understand, analyse and translate concepts and theories into practice. According to Brinkerhoff and Morgan (2010), one of the goals of long-term education programmes is learning to improve understanding and build knowledge and cognitive skills. The researchers believe that long-term education lays the foundation for acquiring theory, concepts and cognitive skills to enhance knowledge for participants to be creative and critical in designing climate change adaptation programmes. At the same time, short-term training guides the implementation and raises awareness of emerging technologies and approaches (Mataya et al., 2020).

2.8 Tools for Identifying and Monitoring Capacity-Building

Research shows that capacity constraints, especially in developing countries, are well documented (Shakya et al., 2018). The scale and uncertainty of the long-term impacts of climate change, its complex and cross-cutting nature, the urgency of action required and the power between different stakeholders mean that managing climate change poses institutional challenges. Therefore, engaging potential champions to co-define entry points for technical interventions and building stakeholders' ownership is urgently required (Shakya et al., 2018).

The tools for identifying and monitoring opportunities for strengthening capacities have been examined by Shakya et al. (2018) and include:

- Regular and broad context assessments engage experts and civil society on changes and opportunities in seven aspects of climate gov-

ernance: evidence, climate policies, awareness, political commitment, participation and influence, institutional capacity and finance.
- The competency framework assesses government skills at the senior and technical levels in mainstreaming climate change (e.g. use of climate information; knowledge of the gender and social implications of interventions), climate finance (e.g. ability to analyse financing needs, access funds and manage their use), strategic communication of climate issues and horizontal (sectoral) and vertical (national to local) coordination.

Research shows that the severity of the climate change impact possesses the capacity to prevent the world from achieving the SDGs. Therefore, it is critical to note that the economic, social and environmental concerns deemed in the 17 SDGs can no longer be dealt with as separate and independent development entities in the context of climate impact (Lozano, 2008; Setó-Pamies & Papaoikonomou, 2016). These goals must be approached in an interdisciplinary and integrated stakeholder management manner, with all stakeholders willing to accept their share of responsibility.

2.8.1 Pitfalls of Government Implementing Climate Actions

Due to the severity of the impact of climate change in developing countries, the haste with which governments implemented projects to build people's resilience led to a reliance on realism that is short term and ad hoc to boost people's capacity. International development organizations, mainly from developed countries, rushed to developing countries to provide one-off training sessions and workshops to boost the capacity of these developing countries. However, such support has a limited impact resulting in low sustainability and institutional capacity in countries to respond to climate change (Williams et al., 2020).

Therefore, there is a need for more and better support and provisions for developing countries to adapt to climate change. The need for such support must be recognized globally, as indicated by the 2015 Paris Climate Change Agreement. The agreement committed developed countries to provide long-term capacity to build countries' resilience to adapt to climate change. Such commitment is enshrined in Article 11 of the 2015 Paris Agreement, which states that capacity-building must operate through appropriate institutional arrangements and be effective, iterative participatory, cross-cutting and gender-responsive (Williams et al., 2020). Consequently, the reality is that the financial resources committed by UNFCC COP21 and COP26 in 2015 and 2021, respectively, are slow in reaching developing countries.

LDCs, SIDS and other developing countries are unreasonably vulnerable to climate change because of their exposure to the impacts of climate change, lack of resources and size constraints (Nurse, 2014; IPCC, 2014a, b, c). Hence, developing countries must take action to identify and integrate local capacity needs into recommendations for policy measures in the United Nations Framework Convention on Climate Change and National Adaptation Planning processes.

2.8.2 Data Collection and Research Methods

Case methods have increasingly become a primary approach researchers and practitioners adopt for data collection. However, the authors triangulate by employing multiple data collection tools (Saunders et al., 2000; Robson, 2002) such as interviews, questionnaires, focus group discussions and participatory action research methodologies (Robson, 2002; Chamber, 2002) to collect data that support case studies. The multi-search methodological approach offers two significant advantages to enhance the quality of the output of the research. These advantages include (a) it provides the researcher with the confidence necessary to pursue the research and (b) it enables triangulation to take place, i.e. the use of different data collection systems within a study to ascertain the data given is the correct information

(Saunders et al., 2000; Robson, 2002). The researcher triangulated sources of information by examining evidence from sources and using that to build a coherent justification for the themes (Creswell, 2009).

Participatory research is a family of approaches, behaviours and methods for enabling people to do their appraisal, analysis, planning, monitoring and evaluation (Chamber, 2002). The methods identified for collecting data are relevant to the research's aims and objectives, the research questions and the target population (Angelsen et al., 2011).

The role of case studies in research can be described as a contradiction, i.e. on the one hand, case studies are widely used by many communities to incorporate research; for example, case study research has consistently been one of the most powerful methods in operations management, particularly in the building of new theory (Dul & Hak, 2008). A case study is anticipated to depict the intricacies of a single case, and the methodology that enables this has developed within the social sciences. The methodology is applied in the social sciences, such as psychology, sociology, anthropology and economics, and in practice-oriented fields, such as environmental studies, social work, education and business studies (Johansson, 2003).

Case studies have been used in varied investigations, particularly in sociological studies, but increasingly in instruction (Stake, 1995). From the perspective of Yin (2014), he viewed the goal of case studies as understanding complex social phenomena and real-life events such as organizational and managerial processes (Yin & Campbell, 1989). The case study is also an ideal methodology when a holistic, in-depth investigation is needed (Feagin et al., 1991). The case is developed in the relationship between the researcher and informants and presented to engage the reader, inviting them to join in this interaction and case discovery.

Researchers utilized a small number of case studies as part of a comparative study, as indicated by Dul and Hak (2008). Dul and Hak (2008) define the case study as a study in which one case (single case study) or a small number of cases (comparative case study) in their real-life context are selected and scores obtained from these cases are analysed qualitatively. The focus on generalization is critical in using case studies as a research method. One of the criticisms of case study research is that the result is not applicable to ordinary life. Case study research is presented as strictly exploratory research in which nothing can be proven, most often by referring to the alleged impossibility of "generalizing" (Dul & Hak, 2008).

As advocated by Yin (2014), the authors utilized the case study methodologies to (a) archival records, (b) interviews, (c) direct observation, (d) use multiple sources of evidence, (e) establish the chain of evidence, (f) have the key informants' review draft study report, (g) do pattern matching or explanation building or time series analysis, (h) use replication logic in multiple case studies, (i) use study protocol and (j) develop case study database tools.

2.9 Top-Down and Bottom-Up Approaches to Enhance the Capacity of Institutions for CCA

Climate change adaptation is often outlined as a community issue (Preston et al., 2013) because climate change impacts are felt mainly at the local level. However, the design, implementation and scaling-up of adaptation strategies and programmes necessitate other stakeholders' knowledge and resources, including government and civil society (Adger et al., 2005; Schipper et al., 2014, in Butler et al., 2015). Many barriers to adaptation occur within the institutional and political environments that govern communities in developing countries (Cannon & Mculler-Mahn, 2010; Ensor, 2011, in Butler et al., 2015). Hence, when instituted, adaptation planning should focus on building the adaptive capacity of all stakeholders to respond to the impact of climate change.

Adaptive capacity is the capability of actors within a structure or system to react to and initiate changes in that structure (Chapin et al., 2006;

Butler et al., 2015). The main features of adaptive capacity are social learning and knowledge exchange, empowerment and connecting social networks that link stakeholders and their resources across administrative levels and spatial scales (Smit & Wandel, 2006; Armitage & Plummer, 2010). The facets of adaptive capacity, when they are integrated, promote knowledge diversity necessary to address complex problems (Folke, 2004); collaborative and inclusive governance, which is spontaneous to change (Armitage & Plummer, 2010); and collective action (Brown, 2008). Consequently, efforts must be made to enhance the adaptive capacity of stakeholders within the communities. This strategy necessitates the commitment of multilevel stakeholders in participatory learning and decision-making actions (Pahl-Wostl, 2009 in Butler et al., 2015). Adaptive capacity processes must be carefully designed and account for stakeholders' power dynamics to ensure their effectiveness (Ballard, 2005). For instance, it is important to remember that the power disparities among the governing elites, the political class and local communities can inhibit neutral and unbiased knowledge exchange and decision-making among the stakeholders (Armitage et al., 2008).

There are two approaches to adaptive capacity planning worth considering in addressing the climate change challenge. These approaches include top-down and bottom-up approaches.

2.9.1 Top-Down Approach

Top-down planning involves government-level decision-making based on long-term regional climate change modelling, impact projections and adaptation strategies identified through technocratic cost-benefit analyses (Kelly & Adger, 2000). This approach has evolved through National Adaptation Programmes of Action to include national and local stakeholders (UNDP, 2010 in Butler et al., 2015).

From the monitoring through observation perspective, the top-down approach is also defined as observing or monitoring efforts demarcated within a global, international or national framework, often focusing on national and international assessments and scientific research.

Top-down approaches define essential variables linked to broad societal benefits and specific agency or operational missions. Therefore, top-down monitoring by large-scale programmes, often informed by high-level frameworks, typically derives system requirements from scientific research programmes or government agency directives at the national or international level (Eicken et al., 2021).

Bottom-Up Approach (Chap. 3)

The JEMS Progressive Community Organization is a non-governmental organization (NGO) working with communities on the southeast of mainland St. Vincent and the Grenadines. The organization has implemented bottom-up approaches to build the capacity and resilience of residents to adapt to the impact of climate change. JEMS has a long history of successfully implementing environmental sustainability programmes. It was awarded the National Youth Award in 1985 and 1986 and the Caribbean Conservation Award in 1993 for its innovative community conservation approach to protect the environment by utilizing cultural art forms, including popular music and poetry, to empower and raise awareness and actions. The work of the organization continues to be stricken by the absence of viable funding to implement its programme. Content within the chapter will cover case studies showcasing examples of the related approach.

2.9.2 Bottom-Up Approach

Bottom-up approaches are observing or monitoring efforts defined and undertaken at the local scale and brought forward to higher-level bodies, often focusing on supporting outcomes desired by a local community. Bottom-up approaches originate at the local community level observing requirements tied to management outcomes that community members or institutions seek to achieve (Danielsen et al., 2009; Commodore et al., 2017; Eicken et al., 2021). Achieving specific goals or management outcomes – particularly considering an environmental change or changing regional human practices – are at the heart of bottom-up efforts.

Top-Down Approach (Chap. 5)

The CYP is the Youth Affairs Division of the Commonwealth Secretariat based in London, UK. It works with governments in 54 member countries of the Commonwealth to build their capacity to utilize top-down approaches to enhance good governance, peace, the rule of law and SDGs. Two-thirds of Commonwealth members are SIDS, which are particularly vulnerable to the impact of climate change. The agency implemented a wide range of development programmes; however, its regional presence allows it to develop "tailor-made" programmes relevant to the needs of youth. The shifting to a hybrid approach enables the agency to achieve successful programme outcomes (forging of the strengths of the top-down and bottom-up approaches). The organization has since dismantled the regional centres. Therefore, it will be necessary for the future to conduct research to assess the impact of the absence of regional centres on the sustainability of its programmes. The content within the chapter will cover case studies showcasing examples of the related approach.

2.9.3 Benefits of Linking Top-Down and Bottom-Up Approaches

It is important to note that both approaches have strengths and weaknesses. For example, top-down planning is sector-based and cannot easily incorporate community priorities (Sherman & Ford, 2013). Research shows that stakeholders from higher-level government and non-governmental organization (NGO) institutions are more aware of long-term global drivers that communities cannot easily conceptualize and can deliver appropriate strategic responses. These stakeholders also have the mandate and resources to create and change policy that enables local-level decision-making (Adger et al., 2005, in Butler et al., 2015). Although community-based PRA strategies can empower marginalized individuals (Chamber, 2002), it is critical to note that a local emphasis fails to incorporate higher-level influences (Scoones, 2009). There is potential to incorporate the strengths of these approaches. The combined approaches could foster greater adaptive capacity than they may achieve when implemented in isolation from each other. Therefore, research shows that linking top-down and bottom-up approaches can complement and fit locally based monitoring most relevant to decision-makers, community planners and large-scale observations. Also, linking these approaches could effectively transition scientist-driven research observations into operational monitoring programmes that inform action and contribute to sustaining observing and monitoring efforts over time (Eicken et al., 2021).

2.9.4 Challenges of Implementing Top-Down and Bottom-Up Approaches

Although studies identify the need to promote collaborations across stakeholders and draw on diverse sources of knowledge (Turnbull et al., 2013), neither of the above research guides on achieving these outcomes or considering how to bridge the potential power dynamics among the key stakeholders. Handbooks and other tools are provided to enhance social learning outcomes, but such tools only target the community level, while the government and other critical stakeholders are excluded from the process (UNDP, 2010). While guidelines for top-down approaches recommend stakeholder analysis and community engagement (UNDP, 2010), a recent review concluded that these processes still prioritize government and expert stakeholders' agendas (Sherman & Ford, 2013; Butler et al., 2015). While adaptation planning provides an opportunity to build the capacity of multiple stakeholders and the various systems they are embedded within, integrating the strengths of top-down and bottom-up approaches to planning and development could generate greater adaptive capacity than either may achieve in separation (Butler et al., 2015). Even though top-down approaches try to include community stakeholders, this has often proved problematic (Sherman & Ford, 2013) because planning guidelines (e.g. UNDP, 2010) do not yet consider design principles that foster social learning, knowledge exchange and power-sharing (Butler et al., 2015).

2.10 Leadership to Enhance Climate Change Adaptation and Resilience

Actions taken today to support climate change adaptation and natural disaster resilience can have far-reaching and positive long-term effects on society and the environment (Allison et al., 2009; Becker & Kretsch, 2019). On the other hand, inaction can set a hazardous and possibly irreversible course affecting numerous facets of human life now and in the future. Research indicates that leadership will be key in developing climate change mitigation and adaptation programmes (Kajsa-Stina Benulic et al., 2021). Therefore, any strategy to tackle climate change underscores the necessity for the transformations of society (European Commission, 2019); that is, the implementation of a comprehensive and nonlinear system changes to make society into something qualitatively different from what it is presently (Feola, 2015; Hölscher et al., 2018; Linnér & Wibeck, 2019). Society is a complex web of systemic structures and processes which can be very difficult to manage. Hence, manipulating transformational change in society would require the development of adaptive structures and responsive mediators and leaders (Linnér & Wibeck, 2019). Over the years, climate change leadership has been combined with environmental leadership. However, climate leadership warrants its ecological niche due to the complexity and uncertainty of climate change and its impact.

Due to the complexity, uncertainty and magnitude of the impact of climate change on society, it will present distinct challenges to individuals wishing to lead climate change and sustainable development-related organizations into the future (Meijerink & Stiller, 2013). It is imperative to note that climate change demands leadership throughout sectorial and governance levels (Torney, 2019; Kajsa-Stina Benulic et al., 2021). Neither leadership literature nor climate literature offers a compromise on the definition of leadership. Both kinds of literature offer examples of other leadership definitions and explain that the leadership concept generally comprises a process, an influence, a group and a common goal. Currently, there is a shortage of research centred on leadership as a barrier to boosting adaptation and building the resilience of people and their communities (Becker & Kretsch, 2019).

One of the major concerns of academia is the evidence of confusion among managers relating to the concept of leadership and management. From a business manager's perspective, Armstrong (1999) defines leadership as a process of inspiring individuals to give their best to achieve the desired result. He contends that it is about getting people to move in the right direction, gaining their commitment and motivating them to achieve their desired goals. Kotle (1996) in Armstrong (1999) advocates the need to separate the concept of management from leadership. He argues that management contains a set of processes, such as budgeting and planning, to ensure that the organization runs smoothly. In contrast, leadership is a set of processes for creating or transforming organizations. It is relevant for today's leaders to be efficient in carrying out such management roles in institutions to ensure effective functioning climate change adaptation actions.

Leaders must occupy the appropriate position within the organization and have the personality and character required to bring the right actors together (Becker & Kretsch, 2019). The leader must work with stakeholders and sectors to create pre- and post-planning fora to enhance the dialogue among key stakeholders and sectors. Since resilience planning is a relatively emerging area in climate change adaptation, such activities allow stakeholders to better understand their roles and risks. Dialogue among stakeholders can clarify the feasible and preferred resilience options (Becker, 2017; Becker & Kretsch, 2019). Such dialogues further instil a sense of ownership in the process and increase trust and participation among stakeholders (Becker, 2017). Organizing workshops and focus groups are a natural fit to bring together the public and private sectors. These activities require little upfront funding and can be conducted in a low-risk, non-threatening manner that engages stakeholders (Becker & Kretsch, 2019).

2.11 Leadership, Gender Mainstreaming and Cooperation for Climate Change Adaptation

Cooperation is needed between different levels of government, policy sectors and public and private agencies to ensure the smooth development and implementation of adaptation policies and programmes. Decision-making in such a situation is often the result of deliberation and negotiation processes in multilevel governance networks (Stiller & Meijerink, 2016). These networks are complex adaptive systems due to (a) the many parties involved and their different perceptions and preferences and (b) the unpredictability of social interactions. The networks are adaptive due to their continuous adjustment to changes in their environment generated by climate change. The individuals and agencies within these networks must continuously cope with numerous uncertainties, comprising the degree of impact of climate change that is anticipated, the swiftness of climate change impact, the regional and local impacts of these changes and the efficiency and effectiveness of adaptation measures to be implemented (Osberghaus et al., 2010).

2.11.1 Leadership

Literature deemed leadership as one of the various considerations that may enhance the adaptive capacity of governance networks (Gupta et al., 2010). This description of leadership goes much further than the traditional hierarchical conception of leadership in which a charismatic, positional leader succeeds in getting supporters for their ideas (Burns, 1978; Stiller & Meijerink, 2016). This notion of leadership is shaped by the complexity leadership theory (CLT), developed in organizational studies (Uhl-Bien et al., 2007). It is portrayed as an integrative framework of leadership functions, including political-administrative, connective, adaptive, enabling and dissemination functions. Its shared purpose is to enhance the adaptive capacity of governance networks (Meijerink & Stiller, 2013). These functions are theoretically associated with the three main leadership challenges these networks faced: trying to increase their adaptive capacity to deal with climate adaptation. The leadership functions encompass (1) influencing the policy process to get adaptation policies accepted and executed; (2) improving connectivity across different policymaking levels, processes, sectors, and actors; and (3) enhancing the capacity of society to learn in response to feedback from the natural system and anticipating long-term impacts of climate change (Stiller & Meijerink, 2016).

Policymakers and practitioners are expected to search for solutions that do not only serve climate change adaptation goals but integrate them into existing policy frameworks. This is known as mainstreaming (Uittenbroek et al., 2013). Stiller and Meijerink (2016) argued that mainstreaming climate adaptation in policy areas such as urban planning, water management and public health implies that stakeholders need to consider the effects of climate change for their respective domains and decide on the implementation of appropriate measures to reduce the vulnerability of their policies to the impacts of climate change. Mainstreaming is understood to inspire policymaking efficiency by combining objectives, increasing the efficient use of human and financial resources and ensuring long-term sustainable investments (Smit & Wandel, 2006). In the local authority's case, horizontal and vertical integration strategies are proposed and applied at the local level to enhance the effectiveness of mainstreaming climate adaptation (Wamsler et al., 2014; Rauken et al., 2015). Vertical top-down strategies such as regulatory, managerial and directed mainstreaming are linked to the implementation process by governmental agencies.

2.11.2 Gender Mainstreaming

On a more basic level, men and women are different in terms of their needs, living conditions and circumstances, including unequal access to and control over power, resources, human rights and institutions, including the justice system, and would be affected differently by the impact of climate change. Gender mainstreaming is a strategy to improve the quality of public policies, programmes and projects to ensure a more efficient allocation of resources. Therefore, better results mean increased well-being for both women and men and the creation of a more socially just and sustainable society. The gender mainstreaming strategy aims to consider these differences when designing, implementing and evaluating policies, programmes and projects so that they benefit both women and men and do not increase inequality but enhance gender equality. Gender mainstreaming aims to solve hidden gender inequalities, providing a tool for achieving gender equality by integrating a gender equality perspective at all stages of the development of climate change adaptation policies, programmes and projects (COE, 2022).

Capacity-building activities and educational materials should be tailored to local needs and contexts.

The authors endorse the recommendations advocated by Bloomfield et al. (2017, p. 21) regarding organizations that are desirous to be involved in capacity-building for climate change adaptation and the SDGs:

- Due to inadequate resources and widespread capacity-building demands, target audiences can be selected by determining what groups could best serve as leaders, trainers or amplifiers of the information. There is great potential for training youth to become those agents of change in their community.
- Organizations interested in capacity-building can benefit from strategic alliances with key partners that play a role in identifying training needs, delivering content, providing expertise, organizing activities and other aspects of capacity-building.
- As an alternative to offering information, courses can have greater outcomes if they are constructed to increase the applicability of the information and allocated to real-life efforts in which participants are involved.
- Providing follow-up support and opportunities to participants who completed training is very important to help them achieve their goals.
- Monitoring efforts should evaluate the capacity-building tools and approaches and engage in adaptive management of the programmes over time.

2.12 Review of Conceptual Frameworks on Capacity-Building for Sustainable Development

Over the past decades, most climate-related capacity-building initiatives have been provided through short-term exercises initiated by bilateral and multilateral agencies. These donors do not invest in systems to enhance the capacity-building of developing countries. Furthermore, donor agencies have no consensus on what capacity-building means or entails (Khan et al., 2019). Although the UNFCCC has emphasized capacity-building as an important part of the intergovernmental process, climate change negotiators have never attempted to officially define it and/or agreed on what percentage of climate finance should be allocated to ensure effective capacity-building takes place in institutions.

Literature on conceptual frameworks for strengthening institutional capacities to build individuals and organizational resilience to adapt to climate change in the future was carefully examined. A conceptual framework offers a comprehensive approach to engaging individuals, organizations and the wider systems that create incentives, i.e. institutions' processes, resources, norms and values (Williams et al., 2020). Research shows that frameworks for social learning in projects and programmes now exist to determine effective strategies and the factors that determine their efficacy (Cundill et al., 2014; Kristjanson et al., 2014; Mataya et al., 2020). The significance of more active learning has also been emphasized as essential to building people's resilience to deal with climate change and other development calamities (Jones et al., 2018). These considerations assist in specifying what adaptation knowledge needs to be shared to build capacity for adaptation.

A conceptual framework laying down the basic principles, priorities, guidelines and scope for capacity-building was adopted by UNFCCC (2001); however, climate change impacts intensified in the two decades that followed. Thus, there needs to be a deliberate injection in human and financial resources at the national level towards building and sustaining the capacity-building needs of key stakeholders' institutions to address the complex nature of the climate challenge (Khan et al., 2019). The authors reviewed several conceptual frameworks to assess their appropriateness and relevance to building the capacity of institutions for climate change adaptation. Here is the analysis of the conceptual frameworks reviewed.

2.12.1 ACT Conceptual Framework for Strengthening Climate Capabilities (Shakya et al., 2018)

The framework was developed through the ACT (Action on Climate Today), implemented in Asia. The ACT is an initiative funded by UK Aid and the UK government and managed by Oxford Policy Management (Shakya et al., 2018). It is an operational framework for strengthening climate capabilities in institutions. The framework was derived from ACT's most successful experiences building institutional capacity and informed by wider literature on governance, climate change and organizational development.

It focuses on building the institutional capacity of government to enhance resilience to climate change in Asia. ACT describes climate change as a wicked problem (Shakya et al., 2018). It sets out pathways to building climate capabilities around the core functions of institutions, including authorizing, resourcing and delivering climate action. These pathways are described through ACT's most compelling case studies for strengthening the institutions. ACT took a flexible and responsive approach to seize political opportunities for change. The approach engages external actors and local experts already skilled in technical analysis. ACT contributed to the building of the capabilities of officials and institutions to support the government's response to climate change and create an informed external constituency for action.

The framework proposes a set of characteristics that institutions should possess to overcome the distinctive challenges in responding to climate change. It states that institutions must be able to act on foresight, learn and adapt, build collaborative coalitions for action, access resources and use incentives to initiate change. These characteristics seek to reframe the perception of capacity-building away from the institutional architecture, policies and plans to one where institutions' actual performance and behaviour are to be respected.

The framework considers how programmes can build capabilities. It introduces the entry points through which capacity-building can be delivered, including the people, organizations, informal norms set by institutions and partners' wider constituencies that shape debates and advise governments. The framework offers a comprehensive approach to thinking through how to engage individuals, organizations and the broader systems that create incentives.

The ACT framework focuses on functional change. The framework's success is measured in terms of the effectiveness of the climate action taken on the ground rather than on the quality of analytical tools used, climate change plans utilized and the number of people trained. The impact delivery has encouraged nuanced, politically astute investment in long-term climate capabilities, not just with the government but with trustworthy supporting partners. The ACT's approach aligns with other global experience that indicates that fostering institutional change can offer high value for money when successful. It can have a long-term impact at a relatively low cost. The ACT framework can be described as a political process with a high risk (Whittle et al., 2012).

The institutional architecture is the bedrock for authorizing, resourcing and delivering climate action in developing countries. The framework emphasizes that institutional capacity should focus on functions rather than forms and structures. It seeks to underscore the value of different design options for functional performance. The framework also unpacks the trade-offs between different designs by utilizing ACT's experience to highlight how understanding any single design's constraints can confirm a consideration for building sufficient accountability to counteract these constraints (Shakya et al., 2018).

2.12.2 Framework for Capacity-Building for Developing Countries (UNFCCC, 2001)

The Seventh Conference of Parties (COP7) to the UNFCCC adopted the Marrakech Accords, which included two frameworks that provide a set of guiding principles and approaches to capacity-building for (a) developing countries and (b) countries with economies in transition (EITs) (Sharma, 2018; UNFCCC, 2001).

The framework's guiding principles on capacity-building for developing countries call for country-driven capacity-building involving learning by doing and building on existing activities. It states that the approach to capacity-building should be continuous, progressive, iterative, effective, efficient, integrated and programmatic. The capacity-building in developing countries should focus on (a) the developing priorities of the developing countries, (b) the unusual conditions of LDCs and SIDSs and (c) marshalling existing national, sub-regional and regional institutions and the private sector to build on existing processes, capacities and capabilities.

The framework has been reviewed under the Subsidiary Body for Implementation (SBI) on four occasions (in 2004, 2007, 2013 and 2019). The framework includes activities such as (a) developing and strengthening skills and knowledge, (b) providing opportunities for stakeholders and organizations to share their experiences and (c) increasing their awareness to enable them to participate more fully in the climate change process. In addition, the framework guides the GEF and other multilateral and bilateral agencies and other intergovernmental organizations and institutions.

The third review of the framework acknowledged areas not covered by the UNFCCC capacity-building framework, including (a) loss and damage associated with climate change impacts and (b) readiness for and access to climate finance. It realized that such information on the finances and severity of the impact of climate change is not often shared by parties in their reports. It also discovered that there is no straightforward process for building on good practices and lessons learned to improve capacity-building efficiency. The reviewers recommend that an institutionalized process ensure appropriate follow-up of lessons learned and good practices.

Other issues observed by the reviewer include (a) capacity-building is less effective when implemented on an ad hoc basis and (b) when it is linked to projects within a limited timeframe and the organization of workshops. The reviewers recognized the need for a structured country-driven approach to create and maintain stable institutions, and foster in-house knowledge across stakeholders, and retain trained personnel and experts to ensure continuity.

2.12.2.1 Gaps and Challenges of the Framework

Among other existing challenges, the review highlighted (a) the lack of financial resources, (b) challenges in setting up permanent institutional arrangements for national reports and greenhouse gas inventories, (c) the need for integrated rather than sectoral training of national experts and (d) the lack of climate-related policy frameworks.

Although the review states a need for more communication from developing countries on the challenges they were facing, countries continue to report persisting gaps in their capacity needs through their National Communications reports submitted to the UNFCCC. The challenges indicated by developing countries were chiefly related to (a) lack of public awareness and support for climate action; (b) lack of training in vulnerability and adaptation assessments and methodologies; (c) inadequate international support for building and retaining individual and institutional capacity; (d) fragmentation of delivery channels, data, experts and research institutions; and (e) lack of solid and permanent institutional arrangements and enabling environments. It is critical to note that these persistent capacity constraints have been attributed to (a) inadequate financial and human resources, (b) lack of ownership and leadership, (c) lack of coordination and fragmentation of capacity-building efforts and (d) ad hoc, short-term, project-based approaches to capacity-building.

The complex and wide-ranging nature of the climate change challenge and the solutions to help address this development challenge makes the matter of capacity challenging to resolve. The proposals from developing countries' parties include an alluring range of requests. Sharma (2018) categorized the capacity development constraints into three important categories, including:

- The capacity to understand the nature of the climate problem that pertains to a country, such as the implications of climate change for various economic and livelihood activities and human and societal well-being.
- The capacity to formulate and implement national actions to limit the scale of the problem through mitigation of greenhouse emissions and to limit the human, ecological, economic and other societal impacts through measures to mitigate risks and adapt to them.
- The capacity to analyse, build consensus on and articulate the national interest in UNFCCC and other climate negotiations and obligations.

The relevant knowledge and skills required to address the climate change challenge in developing countries are still being determined. For example, several LDCs may not be able to run or operate climate modelling endeavours focusing on enhancing the perception of climate change within their respective countries. Therefore, the framework must influence the design of international climate modelling capacity to respond adequately by creating the downscaled scientific information needed by developing countries.

2.12.3 The Adaptation Policy Framework Process (UNDP, 2004)

The *Adaptation Policy Framework Process* (APF) guides studies, projects, planning and policy exercises to identify appropriate adaptation strategies, policies and measures (UNDP, 2004). The APF component is considered a process that helps define current vulnerability and adaptation in the context of the priority system. The APF is a flexible framework depending on the knowledge and vulnerability system level and emphasizes practice rather than theory. The framework begins with the evidence that developing countries already possess concerning vulnerable systems such as agriculture, water resources, public health and disaster management and proposes taking advantage of these synergies and intersecting themes to enable them to reform the policymaking process. According to the UNDP (2004), the APF encompasses five components and is classified as follows:

Component 1: Scoping and designing an adaptation project. This component ensures that a project scale or scope is well integrated into national policy planning and development (Udie et al., 2018). This is the most critical stage of the APF process, and an effective project plan is put in place to implement adaptation strategies, policies and measures.

Component 2: Assessing current vulnerability. The component involves responding to several questions: Where does society stand today regarding vulnerability to climate risks? What factors determine a society's current vulnerability, and how successful are the efforts to adapt to current climate risks?

Component 3: The component assesses future climate risks by developing scenarios of future climate, vulnerability and socio-economic and environmental trends as a basis for considering future climate risks.

Component 4: This concerns the formulation of an adaptation strategy in response to immediate and future climate risks and requires the identification and selection of a set of adaptation policy options and measures and the formulation of these options into a cohesive, integrated strategy.

Component 5: The continuation of the adaptation process, such as implementing, monitoring, evaluating, improving and sustaining the initiatives launched by the adaptation project.

Two cross-cutting processes complement the components articulated in the APF. These are (a) engaging stakeholders in the adaptation process and (b) assessing and enhancing the adaptive capacity of the area. The engagement of stakeholders to participate in the adaptation process is an essential component of the APF. This is necessary to sustain dialogue among stakeholders. While assessing and enhancing adaptive capacity requires catalysing change management processes so that societies can better cope with climate change, the process can be deemed time- and resource-consuming and is manageable when properly implemented.

The APF recognizes the value of building on what is already known, utilizing synergies and intersecting themes to enable more informed and effective policymaking to guide the adaptation process. The APF placed emphasis on flexibility. When implementing the framework, stakeholders are encouraged only to use those components of the conceptual framework that meet their unique needs.

2.12.4 Sustainable Livelihoods Framework (DfID, 1999)

The Department for International Development (DfID) developed the Sustainable Livelihoods Framework (SLF) in 1999 to provide a tool for understanding community livelihoods, vulnerabilities and outcomes of community livelihood strategies in specific or various sectors of society (DfID, 1999, 2000, in Angelsen et al., 2011). SLF is a framework for gathering contextual information. The underlying message conveyed through the framework is constructed by various forces, including shocks, trends, reasonability and factors such as transforming structures and processes. These processes and factors exist within a state that is constantly changing. The SLF consists of five components: vulnerability context, livelihood assets, transforming structures and processes, livelihood strategies and livelihood outcomes. The assessment of livelihoods necessitates a continuous assessment of the assets available within the community, the livelihood outcomes that the community is hoping to accomplish and the livelihood strategies which the community is adapting with the hope of achieving the designated outcomes at a later stage in the development process.

Although the framework is silent on specific strategies for building the institutional capacity to adapt to climate change in developing countries, it has relevance when dealing with climate change impact – shocks and vulnerabilities. One component of the framework focuses on addressing the vulnerability caused by forces such as shocks, including health shocks, economic shocks, crops, live stocks failure shocks and others (Angelsen et al., 2011). These shocks can also be caused by climate change and other calamities in developing countries (Sem, 2007; Nurse et al., 2014; Hernandez-Delgado, 2015). Trends can be caused and driven by population growth, resource use, economics, governance and technological trends, and seasonality can be influenced by trends. Understanding the vulnerability context would provide the following benefits: an understanding of adaptive capability, coping strategies and the role of the community's environmental and natural resource base in securing livelihoods and producing safety nets for the population.

The SLF livelihood assets are based on a belief that human beings require a set of assets or capital. These capitals include human, social, physical, financial and natural capital to achieve their livelihood outcomes. These assets have relevance to the achievement of sustainable development. The formation of structures is an important aspect of the SLF. This includes government, the private sector and processes, focusing on laws, policies, culture and institutions. These structural factors are significant in enabling the community or society to access livelihood assets. Another component of the SLF is the mechanism, which permits the community to adapt to achieve the desired livelihood outcomes. The achievement of livelihood outcomes will depend on the effectiveness of the community in combining (a) transforming structures and processes and (b) the livelihood assets into innovative strategies to allow the community to adapt to the vulnerability. Although the SLF is not explicitly designed to address capacity-building, the authors use components from the SLF framework to develop a new conceptual framework on capacity-building for climate change adaptation in developing countries.

After careful analysis of the conceptual frameworks reviewed, the authors realized that none of the frameworks analysed in their entirety was appropriate as a stand-alone conceptual framework for the study. However, some of the elements from the various frameworks can be chosen to contribute to developing a new conceptual framework on capacity-building for climate change adaptation in developing countries. Table 2.2 presents some of the elements taken from the various conceptual frameworks.

2.13 New Conceptual Framework for Climate Change Adaptation Capacity-Building in Developing Countries

Borrowing components from the various frameworks reviewed, a new conceptual framework relevant to the capacity-building needs in developing countries has been constructed. Hence, the new conceptual framework is entitled Climate Change Adaptation Framework for Capacity-Building in Developing Countries (Fig. 2.1).

The framework consists of four steps; each consists of a set of strategic tools to operationalize the framework. The steps of the Climate Change Adaptation Framework for Capacity-Building in Developing Countries are as follows:

Step 1: Climate Risk Vulnerability Context

Step 1 of the framework is adapted from the DfID (1999 in Angelson, 2011) Sustainable Livelihoods Framework (SLF) and Simmons (2021) Youth Climate Change Adaptation Framework. It is framed in terms of vulnerability because of the climate change stressors impacting developing countries and as a framework for learning to put theory into a practical perspective (DfID, 1999 in Angelson, 2011; Park, 2012; Wise et al., 2014; Simmons, 2021). The climate change pressures and stressors identified in the figure are the major triggers of vulnerability to human beings and ecosystem systems. According to the literature, vulnerability is a contextual issue and depends on the adaptive capacity of the system in consideration (IPCC, 2014a, b, c, 2022; Nurse, 2014).

Notably, the possibility of identifying vulnerability will be contingent on the context of the climate change situation being researched in developing countries (Nurse et al., 2014; UNEP, 2013). Consequently, it is necessary to understand the context of the researched area. The first stage is focused on developing adaptive strategies

Table 2.2 Elements used from each conceptual framework to be incorporated into a new framework

Conceptual Framework 1: Framework for strengthening climate change capabilities	Conceptual Framework 2: Framework for capacity-building for developing countries (UNFCCC, 2001)	Conceptual Framework 3: The Adaptation Policy Framework Process (UNDP, 2004)	Sustainable Livelihoods Framework (DfID, 1999 in Angelsen et al., 2011)
Setting out pathways to build climate change capabilities is a core function of institutions, including authorizing, resourcing and delivering climate actions Flexible and responsive approach that seizes political opportunities for change that emerges Engaging external actors and local experts who are already credible in undertaking technical analysis Identifying entry points for capacity-building such as people, organization, informal norms set by organizations and broader constituency of partners	Structure a country-driven approach to create and maintain stable institutions, secure and foster in-house knowledge across stakeholders and retain trained persons and experts to ensure continuity An institutionalized process be instituted to ensure appropriate follow-up of lessons learned and good practices Lobby the international community for inclusion of loss and damage associated with climate change impacts Enhance the capacity of the government for readiness for and access to climate finance Building on good practices and lessons learned to improve the efficiency of capacity-building Develop a country-driven approach to create and maintain stable institutions, secure and foster in-house knowledge across stakeholders and retain trained personnel and experts to ensure continuity	Development and strengthening of skills and knowledge Providing opportunities for stakeholders and organizations to share their experiences Increase their awareness to enable them to participate more fully in the climate change process Components 1, 2 and 3 Placed emphasis on flexibility Encouraged only to use those techniques and components that meet their unique needs when implementing the framework Recognizes the value of building on what is already known, utilizing synergies and intersecting themes to enable more informed and effective policymaking to guide the adaptation process	Tool for understanding community livelihoods, vulnerabilities and outcomes of community livelihood strategies Five interacting components, namely, the vulnerability context, livelihood assets, transforming structures and processes, livelihood strategies and livelihood outcomes Addressing the vulnerability, which is caused by forces such as shocks which include health shocks, economic shocks, crops and live stocks failure shocks Based on a belief that human beings require a set of assets or capitals (human, social, physical, financial and natural capitals) to achieve their livelihood outcomes and/or sustainable development Formation of structures that are significant in enabling the community or society to access livelihood assets The achievement of livelihood outcomes will depend on the effectiveness of the community in combining (a) transforming structures and processes and (b) the livelihood assets into innovative strategies to allow the community to adapt to the vulnerability

to raise awareness of the climate stressors impacting developing countries and identify strategies for building people's resilience to adapt to climate change. This stage utilized top-down and bottom-up approaches to assess the current vulnerabilities brought about by the impact of the climate stressors. It will also assess future risks resulting from the exposure of an area to the continuous impact of climate change in the future (UNDP, 2004).

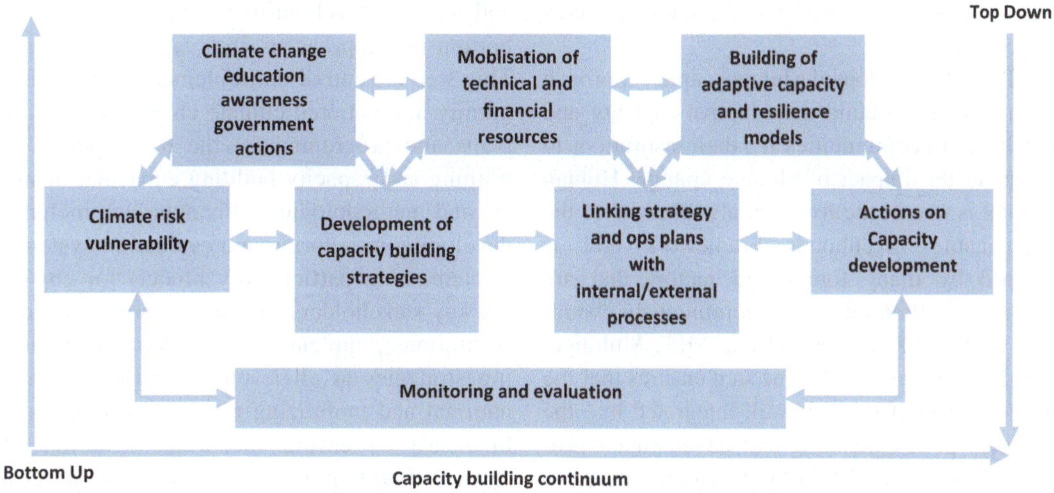

Fig. 2.1 Climate change adaptation framework for capacity-building in developing countries

The assessment process includes sectorial levels of socio-economic impacts, such as the impact of climate stressors on agriculture, tourism, fisheries and other forms of livelihoods and sectors, as well as the impact of climate stressors on the environment, including forestry, coral reef, mangrove and water supply, and on future trends. In essence, step 1 of the conceptual framework focuses on stakeholder engagement to scope the vulnerabilities and possible adaptations existing in the institution or community (depending on the application area) (Udie et al., 2018).

The government and other key stakeholders later in the framework could utilize the information from the assessment to collaborate with civil society and community-based organizations (CBOs) to formulate community and national climate change adaptation plans and programmes. An important component of this step is the development of a capacity-building initiative. The authors utilized multi-research methods to collect data for this step of the conceptual framework, including policymakers and other key stakeholders, including youth and women. This included review of socio-economic data and statistics, utilizing data collection tools such as interviews, questionnaires, focus group discussions and risk mapping exercises to collect data. Participatory Climate Risk Vulnerability Mapping exercises allowed the participants to validate the climate pressures identified through a literature review via stakeholders' engagement (Simmons, 2018, 2021). A climate vulnerability report will be prepared after the data analysis exercise. These reports would be used to prepare the case study.

Step 2: The Development of Capacity-Building Strategies

Step 2 of the conceptual framework focuses on the interpretation and analysis of the data collected on the nature and extent of the vulnerabilities and decides on the appropriate capacity-building approach to responding to the vulnerabilities faced by developing countries. Based on the data analysis, deciding whether a top-down, bottom-up or hybrid (mix of both) approach would be most appropriate to resolve the situation is critical. The authors adapted Calgaro's (2011) concept of shock and stressors and DfID's study (1999, in Angelsen et al., 2011), which theorized that livelihood is impacted by trends and shocks and seasonality over which people have limited control. These shocks include human, economic, natural and crop/livestock health. In terms of seasonality, these may include prices for basic products, health relating to the impact of invasive species and diseases and natural resources and diversity, which may be destroyed due to the impact of drought, storms and hurricanes on biodiversity, or the impact of sea level rise and higher sea surface temperatures on coral reefs.

The authors adopted a human capital approach as a strategy to build the adaptive capacity and resilience of communities and their institutions to adapt to the impact of climate change. Human capital is one of the five capitals or assets of the Sustainable Livelihoods Framework and is shaped by many forces and factors that are dynamic in themselves and continuously changing (DfID, 2000, in Angelsen, 2011; Mulongo, 2012; Simmons, 2018). The step ensures that the project scale or scope is well integrated into the national policy planning and development process (Udie et al., 2018) by developing adaptive policy options and measures and formulating coherent, integrated strategies to enhance climate change adaptation. An important part of step 2 is linking the capacity-building initiative with regional and international climate change adaptation and mitigation policies and programmes, including actions of CCCCC (Caribbean) and UNFCCC negotiations (international). Identifying key stakeholders and partners to assist the resource mobilization process in financing the climate change adaptation and mitigation actions at national and community levels is critical to achieving the climate change adaptation outcomes. This strategic approach ensures that human, financial and material resources are available to implement the community and national climate change adaptation plans and programmes successfully.

The authors utilize a multi-research methodology to collect data for this conceptual framework stage from policymakers and other key stakeholders, including youth and women. The review of the developing countries' human, institutional and resource capacities would be implemented through statistical data analysis, development reports, interviews, questionnaires and focus group discussions where possible.

Step 3: Building Capacity Through Linking Strategic and Operational Plans with Internal and External Partners

Step 3 of the conceptual framework concentrates on implementing the adaptive capacity-building process and programmes at community and national levels utilizing the top-down and bottom-up approaches to capacity-building. This step is centred on implementing the community and national climate change adaptation plans and programmes – the development of training and capacity-building curricula, manuals and methodologies. This approach includes developing monitoring and evaluation systems, implementing participatory models for engaging key stakeholders (people, communities and institutions), implementing gender mainstreaming strategies at all levels of institutions and national and mobilizing resources (human and financial) to ensure the implementation of capacity-building strategies and programmes. An important part of this strategy is to ensure that adequate resources such as experts, materials and finances are mobilized to implement the adaptive climate change processes efficiently (UNDP, 2004).

The agency/agencies would make a concerted effort to mobilize external resources from the Green Climate Fund, UNFCCC and bilateral and multilateral agencies to provide technical support and financial resources to implement adaptive capacity-building initiatives of relevant institutions to implement national climate change adaptation plans and projects.

This will provide opportunities for policymakers and practitioners to search for solutions that do not only serve climate adaptation goals but integrate and mainstream them in existing national policy planning processes and frameworks. The mainstreaming (Uittenbroek et al., 2013; Stiller & Meijerink, 2016) stimulates the effectiveness of policymaking by combining objectives, increasing the efficient use of human and financial resources and ensuring long-term sustainable investments (Smit & Wandel, 2006). Gender mainstreaming will be a critical component of the capacity development mainstreaming process.

A multi-research approach would be utilized to collect data on the performance of stage 3. Reports will be reviewed on project-implemented processes.

Step 4: Capacity Development Initiatives Instituted

Step 4 of the conceptual framework centres on implementing training and other capacity-building strategies to raise awareness of the population's impact and risk associated with climate change and devise adaptation projects/programmes to adapt to the impacts. Capacity-building is not a short-term approach; therefore, practical education and training, for instance, cannot be achieved by implementing one-of-a-kind or one-off training activities. Continuous education and training are essential for appropriate capacity-building at community and national levels (United Nations, 2015). Climate change has demonstrated that it is no longer effective to train people to meet their current needs but to develop skills and knowledge that support entrepreneurship, resilience and innovations in enterprise development and their transition to sustainable practices (ILO, 2013; Poschen, 2015). Government and training providers are responsible for ensuring continued review and access to training programmes that support lifelong skills development and focus on future planning and development needs (ILO, 2010; Poschen, 2015). It is also necessary for people to be constantly engaged in developing and implementing climate change adaptation projects/programmes and constantly learn from these initiatives.

This step of the conceptual framework links capacity-building approaches with the community and national climate change adaptation plans and programmes formulated in step 3. In this step, organizers would implement participatory top-down, bottom-up and/or hybrid approaches to engage people, organizations and their constituencies to participate in the overall development and implementation of the climate change adaptation initiative(s). The overall purpose of this step is to provide opportunities for the organizers to analyse, build consensus and articulate national, regional and international positions regarding climate change negotiation in collaboration with UNFCCC and other appropriate institutions. Although the framework is designed for the successful implementation of SDG17, capacity development, the framework is also applicable for achieving SDG13 on climate change, specifically in developing countries and SIDS.

The authors utilized a multi-research approach to collect data on the performance of stage 4 and reported on the performance of the capacity-building initiatives instituted and their effectiveness in developing and sustaining adaptive capacity and building the resilience of the community and national constituencies to adapt to the impact of climate change.

2.14 Summary of Chapter

Local governance and climate change adaptation initiatives suffer from issues inherent to the development of SIDS and LDCs, such as (a) insufficient capacity resulting from lack of technical know-how, (b) limited financial and human resources, (c) stringent legislation and policy frameworks and (d) ineffective monitoring and evaluation mechanisms which frustrate the implementation of effective climate change adaptation initiatives (Williams et al., 2020). The concept of capacity-building has historically been constructed vaguely, such that many possible meanings coexist under the single heading of capacity-building. The effectiveness of sustainable capacity-building initiatives is plagued with structural challenges, such as the short-term focus, weak, project-based, consultancy-led and donor-driven. Therefore, to overcome the challenges associated with the current capacity-building programmes, a new partnership model between the funders and the recipients' countries should be introduced to address the capacity-building challenges developing countries encounter (Khan et al., 2019). Developing countries must ensure that capacity-building is addressed during the UNFCCC negotiations as a long-term development priority. The agreed and necessary technical and financial resources allocated to support such an initiative should be timely delivered in association with a transparent implementation feedback mechanism.

The knowledge and information required to tackle climate change and the implementation of relevant UN Sustainable Development Goals

(SDGs) should be viewed as global public goods and must be shared by developed and developing countries. It is also critical to note that the necessary human and financial resources should be available to implement a new vibrant and viable conceptual framework on capacity-building for climate change adaptation in developing countries.

Vulnerable countries must have rich indigenous knowledge based on age-old experiential learning, experience and practices about implementing climate change adaptation, and such knowledge is integrated into the design and development of the new framework. All stakeholders should buy into this viable new conceptual framework on Climate Change Adaptation Framework for Capacity-Building in Developing Countries. This development model should be shared for mutual advantage by all parties, supported by the mobilization of the necessary finances and financial mechanisms and channelled to the recipient country to support the capacity-building process (Khan et al., 2019).

References

Adger, W. (2003). Social capital, collective action, and adaptation to climate change. *Economic Geography, 79*(4), 387–404. Available at: http://link.springer.com/chapter/10.1007/978-353192258-4_19

Allison, I., Bindoff, N. L., Bindschadler, R. A., Cox, P. M., de Noblet, N., England, M. H., Francis, J. E., Gruber, N., Haywood, A. M., Karoly, G., Kaser, D. J., Le Quéré, C., Lenton, T. M., Mann, M. E., McNeil, B. I., Pitman, A. J., Rahmstorf, S., Rignot, E., Schellnhuber, H. J., Schneider, S. H., Sherwood, S. C., Somerville, R. C. J., Steffen, K. Steig, E. J., Visbeck, M., & Weaver, A. J. (2009). *The Copenhagen Diagnosis, 2009: Updating the World on the Latest Climate Science*. The University of New South Wales Climate Change Research Centre (CCRC), Sydney, Australia.

Angelsen, A., Larsen, H. O., Lund, J. F., Smith-Hall, C., & Wunder, S. (2011). *Measuring livelihoods and environmental dependence; methods for research and fieldwork.* Earthscan.

Armitage, D., & Plummer, R. (Eds.). (2010). *Adaptive capacity and environmental governance*. Springer.

Armitage, D., Marschke, M., & Plummer, R. (2008). Adaptive co-management and the paradox of learning. *Global Environmental Change, 18*, 86–98.

Armstrong, M. (1999). *How to be an even better manager* (5th ed.). Kogan Page Limited.

Ballard, D. (2005). Using learning processes to promote change for sustainable development. *Action Research, 3*(2), 135–156.

Becker, A. (2017). Using boundary objects to stimulate transformational thinking: Storm resilience for the port of Providence, Rhode Island (USA). *Sustainability Science, 12*, 477–501. https://doi.org/10.1007/s11625-016-0416-y

Becker, A., & Kretsch, E. (2019). The leadership void for climate adaptation planning: Case study of the Port of Providence (Rhode Island, United States). *Frontiers in Earth Science, 7*, 29. https://doi.org/10.3389/feart.2019.00029

Bergeron, K., Abdi, S., DeCorby, K., Mensah, G., Rempel, B., & Manson, H. (2017). Theories models and frameworks used in capacity building interventions relevant to public health: a systematic review. *BMC Public Health, 17*(1) 10.1186/s12889-017-4919-y

Bloomfield, G., Bucht, K., Martínez-Hernández, J. C., Ramírez-Soto, A. F., Cheseña-Hernández, I., Lucio-Palacio, C. R., & Inzunza, E. R. (2017). Capacity-building to advance the United Nations sustainable development goals: An overview of tools and approaches related to sustainable land management. *Journal of Sustainable Forestry*. https://doi.org/10.1080/10549811.2017.1359097

Bolger, J. (2000). Capacity development: Why, what and how. In *Capacity development.*, Occasional series (Vol. 1(1)). CIDA.

Boyd, E., Ensor, J., Broto, V. C., & Juhola, S. (2014). Environmentalities of urban climate governance in Maputo, Mozambique. *Global Environmental Change, 26*(1), 140–151. https://doi.org/10.1016/j.gloenvcha.2014.03.012

Brinkerhoff, D. W., & Morgan, W. P. J. (2010). Capacity and capacity development: Coping with complexity. *Public Administration and Development, 10*, 2–10. https://doi.org/10.1002/pad

Brown, V. A. (2008). *Leonardo's vision: A guide to collective thinking and action*. Sense Publishers.

Burns, J. M. (1978). *Leadership*. Harper & Row.

Butler, J. R. A., Wise, R. M., Skewes, T. D., Bohensky, E. L., Peterson, N., Suadnya, W., Yanuartati, Y., Handayani, T., Habibi, P., Puspadi, K., Bou, N., Vaghelo, D., & Rochester, W. (2015). Integrating top-down and bottom-up adaptation planning to build adaptive capacity: A structured learning approach. *Coastal Management, 43*, 346–364. Taylor & Francis Group, LLC. ISSN: 0892-0753 print /1521-0421 online. https://doi.org/10.1080/08920753.2015.1046802

Calgaro, E. (2011). *Building resilient tourism destination futures in a world of uncertainty: Assessing destination vulnerability in Khao Lak, Patong, and Phi Don, Thailand to the 2004 Tsunami*. Macquarie University Research Online.

Callaghan, I. (2015). Climate Finance after COP21: Pathway to effective Financing of Commitments and Needs: Invest Watch: Ian Callaghan Associates Limited 2015 Final 011215_MG2.

Chamber, R. (2002). *Participatory workshops: A sourcebook of 21 sets of ideas and activities* (Vol. 794, pp. 421–434). Earthscan.

Chapin, F. S., Lovecraft, A. L., Zavaleta, E. S., Nelson, J., Robards, M. D., Kofinas, G. P., Trainor, S. F., Peterson, G. D., Huntingdon, H. P., & Naylor, R. L. (2006). Policy strategies to address sustainability of Alaskan boreal forests in response to a directionally changing climate. *Proceedings of the National Academy of Sciences of the United States of America, 103*, 16637–16643.

Chevallier, R. (2010). Integrating adaptation into development strategies: The Southern African perspective. *Climate and Development, 2*(2), 191–200. https://doi.org/10.3763/cdev.2010.0039

CHF. (2008, February). *Capacity development: Key to North-South NGO partnerships? CHF partners in rural development*. Available at: www.chfpartners.ca/publications/documents/CapacityBuildingCaseStudyFinalFeb408.pdf. Accessed 1 Sept 2009.

COE (2022). https://www.coe.int/en/web/genderequality/what-is-gender-mainstreaming.

Commodore, A., Wilson, S., Muhammad, O., Svendsen, E., & Pearce, J. (2017). Community-based participatory research for the study of air pollution: A review of motivations, approaches, and outcomes. *Environmental Monitoring and Assessment, 189*, art. 378.

Connolly, P., & York, P. (2002). Evaluating capacity-building efforts for nonprofit organisations. *OD Practitioner, 34*, 33–39.

Corfee-Morlot, J., Cochran, I., Hallegatte, S., & Teasdale, P. J. (2011). Multilevel risk governance and urban adaptation policy. *Climatic Change, 104*(1), 169–197.

Creswell, J. W. (2009). *Research Design, Qualitative, Quantitative and Mixed Methods Approaches*, (3rd Ed.), London: Sage.

Cundill, G., Shackleton, S., Sisitka, L., Ntshudu, M., Lotz-Sisitka, H., Kulundu, I., & Hamer, N. (2014). *Social learning for adaptation: A descriptive handbook for practitioners and action researchers*. IDRC/Rhodes University/Ruliv.

Danielsen, F., et al. (2009). Local participation in natural resource monitoring: A characterization of approaches. *Conservation Biology, 23*, 31–42.

Dany Va, Bowens, K. J., & Miller, F. (2015). Assessing the institutional capacity to adapt to climate change: A case study in the Cambodian health and water sectors. *Climate Policy, 15*(3), 388–409. https://doi.org/10.1080/146930

DFID (1999). *Sustainable livelihoods guidance sheets 1–2*. DFID, London.

DFID (2000). *Sustainable livelihoods guidance sheets. DFID*, London.

Di Gregorio, M., Fatorelli, L., Paavola, J., Locatelli, B., Pramova, E., Nurrochmat, D. R., May, P.,H., Brockhaus, M., Sari, I. M.,& Kusumadewi, S. D. (2019). Multi-level governance and power in climate change policy networks. *Global Environmental Change, 54*, 64–77

Dul, J., & Hak, T. (2008). *Case study methodology in business research*. Routledge.

Enemark, S. (2003, April). Understanding the concept of capacity-building and the nature of land administration systems. *FIG Working Week*.

England, M. I., Dougill, A. J., Stringer, L. C., Vincent, K. E., Pardoe, J., Kalaba, F. K., et al. (2018). Climate change adaptation and cross-sectoral policy coherence in Southern Africa. *Regional Environmental Change, 18*(7), 2059–2071. https://doi.org/10.1007/s10113-018-1283-0

Ensor, J., & Harvey, B. (2015). Social learning and climate change adaptation. Evidence for international development practice. *WIREs Climate Change, 6*(5), 509–522. https://doi.org/10.1002/wcc.348

Eicken, H., Danielsen, F., Sam, J. M., Fidel, M., Johnson, N., Poulsen, M. K., Lee, O. A., Spellman, K. V., Iversen, L., Pulsifer, P., & Enghoff, M. (2021). Connecting Top-Down and Bottom-Up Approaches in Environmental Observing, *BioScience, 71*(5) 467–483.

Etongo, D. & Gill, K. (2022). Local governance capacity needs for implementing climate change adaptation in Seychelles: An assessment based on the capital approach. *Challenges, 13*(2):49. https://doi.org/10.3390/challe13020049

European Commission. (2019). Communication from the Commission to the European Parliament, the European Council, the Council, The European Economic and Social Committee and the Committee of the Regions. The European Green Deal, COM. (2019)*640*, final.

Feagin, J. R., Orum, A. M., & Sjoberg, G. (1991). *A case for the case study*. UNC Press Books.

Feola, G. (2015). Societal transformation in response to global environmental change: A review of emerging concepts. *Ambio, 44*(5), 376–390. https://doi.org/10.1007/s13280-014-0582-z

Folke, C. (2004). Traditional knowledge in social-ecological systems. *Ecology and Society, 9*(3), 7.

Freeman, K. (2010). *Capacity development theory and practice*. Master's thesis, Oxford Brooks University.

Gordon, J. & Chadwick, K. (2007) *Impact assessment of capacity building and training: assessment framework and two case studies*. Impact Assessment Series Report No. 44, February 2007. https://iaes.cgiar.org/sites/default/files/pdf/76.pdf

Gupta, J., Termeer, K., Klostermann, J., Meijerink, S., van den Brink, M., Jong, P., Nooteboom, S., & Bergsma, E. (2010). The adaptive capacity wheel: A method to assess the inherent characteristics of institutions to enable the adaptive capacity of society. *Environmental Science & Policy, 13*, 459–471. https://doi.org/10.1016/j.envsci.2010.05.006

Hernandez-Delgado, E. A. (2015). *The emerging threats of climate change on tropical coastal ecosystem service, public health, local economies, and livelihoods sustainability of small islands: Cumulative impacts and synergies*. Marine Pollution Bulletin, University of Puerto Rico, Centre for Applied Tropical Ecology

and Conservation, Coral Reef Research Group Puerto Rico, USA.
Hölscher, J. K., Wittmayer, J. M., & Loorbach, D. (2018). Transition versus transformation: What's the difference? *Environmental Innovation and Societal Transitions, 27*, 1–3. https://doi.org/10.1016/j.eist.2017.10.007
https://sdgs.un.org/topics/capacity-development
https://www.coe.int/en/web/genderequality/what-is-gender-mainstreaming
ILO. (2010, November). *A skilled workforce for sustainable growth: A G20 Training Strategy*. International Labour Office.
ILO. (2013). *Sustainable development, decent work and green jobs*. International Labour Conference, 102nd Session, International Labour Office.
IPCC. (2014a). Climate change (2014): Synthesis report. In *Contribution of working groups I, II and III to the fifth assessment report of the Intergovernmental Panel on Climate change* (p. 151). IPCC.
IPCC. (2014b). Summary for policymakers. Climate change impacts, adaptation, and vulnerability. In C. B. Barros, V. R. Dokken, D. Mach, K. J. Mastrandrea, M. D. Bilir, T. E. Chatterjee, M. Ebi, & K. L. Estrada (Eds.), *Contributions of the working group II to the fifth assessment report, Field*.
IPCC. (2014c). Climate change: Synthesis report. In *Contribution of working group 1, 11, 111 to the fifth assessment report of the Intergovernmental Panel on Climate change* (p. 151). IPCC.
IPCC. (2022). *Climate change 2022 mitigation of climate change, summary for policy makers: Working group 111 contribution to the sixth assessment report of the International Panel on Climate change* (AR6 WG 111). IPCC.
Jayatilaka, W. (2003). Capacity-building for sustainable development. *Journal of the National Science Foundation of Sri Lanka, 31*, 81–97.
Johansson, R. (2003). Case study methodology. In *Methodologies in Housing Research, Stockholm*.
Jones, L., Harvey, B., Cochrane, L., Cantin, B., Conway, D., Cornforth, R. J., et al. (2018). Designing the next generation of climate adaptation research for development. *Regional Environmental Change, 18*, 297–304.
Kajsa-Stina Benulic, Kropf, M., Linnér, B.-O., & Wibeck, V. (2021). The meaning of leadership in polycentric climate action. *Environmental Politics*. https://doi.org/10.1080/09644016.2021.1970087
Kelly, P. M., & Adger, W. N. (2000). Theory and practice in assessing vulnerability to climate change and facilitating adaptation. *Climatic Change, 47*, 325–352.
Khan, M., Mfitumukiza, D., & Huq. S. (2019). Capacity-building for implementation of nationally determined contributions under the Paris Agreement. *Climate Policy*. https://doi.org/10.1080/14693062.2019.1675577
Kristjanson, P., Harvey, B., Van Epp, M., & Thornton, P. K. (2014). Social learning and sustainable development. *Nature Climate Change, 4*, 5–7.
Kuhl, S. (2009). Capacity development as the model for development aid organisations. *Development and Change, 40*(3), 551–557.
Lafontaine, A. (2000). *Assessment of capacity development efforts of other development cooperation agencies*. GEF-UNDP.
Leck, H., & Simon, D. (2013). Fostering Multiscalar Collaboration and Co-operation for Effective Governance of Climate Change Adaptation. *Urban Studies, 50*(6), 1221–1238.
Lozano, R. (2008). Envisioning sustainability three-dimensionally. *Journal of Cleaner Production, 16*(17), 1838–1846.
Linnér, B.-O., & Wibeck, V. (2019). *Sustainability transformations: Drivers and agents across societies*. Cambridge University Press.
Lopes, C., & Theisohn, T. (2003). *Ownership, leadership, and transformation: Can we do better for capacity development?* Earthscan.
Lusthaus, C., Adrien, M. H., Anderson, G., & Garden, F. (1999). *Enhancing organizational performance a toolbox for self-assessment*. Ottawa, Canada: International Development Research Centre. Retrieved from https://www.idrc.ca/en/book/enhancing-organizational-performance-toolbox-self-assessment
Mataya, D.C., Vincent, K., & Dougill, A. J. (2020). How can we effectively build capacity to adapt to climate change? Insights from Malawi. *Climate and Development, 12*(9), 781–790. Francis and Taylor Group. https://doi.org/10.1080/17565529.2019.1694480
Meijerink, S., & Stiller, S. (2013). What kind of leadership do we need for climate change adaptation? A framework for analyzing leadership objectives, functions and tasks in climate change adaptation. *Environment and Planning C, 31*, 240–256. https://doi.org/10.1068/c11129
North, D. C. (1990). *Institutions, Institutional Change and Economic Performance*. Cambridge: Cambridge University Press.
Mulongo, G. (2012). *Human Capital in Education: Principles, critiques and current thinking*. Institute of Education, University of London.
Nurse, L. A., McLean, R. F., Agard, J., Briguglio, L. P., Duvat-Magnan, V., Pelesikoti, N., Tompkins, E., & Webb, A. (2014). Small islands. In V. R. Barros, C. B. Field, D. J. Dokken, M. D. Mastrandrea, K. J. Mach, T. E. Bilir, M. Chatterjee, K. L. Ebi, Y. O. Estrada, R. C. Genova, B. Girma, E. S. Kissel, A. N. Levy, S. MacCracken, P. R. Mastrandrea, & L. L. White (Eds.), *Climate change 2014: Impacts, adaptation, and vulnerability. Part B: Regional aspects. Contribution of working group II to the fifth assessment report of the Intergovernmental Panel on Climate change* (pp. 1613–1654). Cambridge University Press.
OECD DAC. (2019). Better evaluation, revised evaluation criteria. Definition and principles for USE. OECD/DAC Network on Development Evaluation. Revised evaluation criteria Dec 2019. Pdf.

Ojwang, L., Rosendo, S., Celliers, L., Obura, D., Muiti, A., Kamula, J., & Mwangi, M. (2017). Assessment of coastal governance for climate change adaptation in Kenya. *Earth's Future, 5*, 1119–1132. https://doi.org/10.1002/2017EF000595

Organisation for Economic Co-operation and Development (OECD). (2006). *The challenge of capacity development: Working towards good practice* (DAC guidelines and reference series) (p. 46). OECD.

Osberghaus, D., Dannenberg, A., Mennel, T., & Sturm, B. (2010). The Role of the Government in Adaptation to Climate Change. *Environment and Planning C: Government and Policy, 28*(5), 834–850.

Pahl-Wostl, C., Conca, K., Kramer, A., Maestu, J., & Schmidt, F. (2013). Missing links in global water governance: A processes-oriented analysis. *Ecology and Society, 18*(2), 33. https://doi.org/10.5751/ES-05554-180233

Park, J. J. (2012). Fostering community energy and equal opportunities between communities. *Local Environment, 17*(4), 387–408. Available at: http://www.tandfonline.com/doi/abs/10.1080/13549839.2012.678321. Accessed 2 Apr 2022.

Pearson, J. (2011a). *The core concept, Part I*. Available at: www.Lencd.org/learning

Poschen, P. (2015). *Decent work, green jobs and the sustainable economy: Solution for climate change development*. International Labour Organisation.

Preston, B., Mustelin, J., & Maloney, M. (2013). Climate adaptation heuristics and the science/policy divide. *Mitigation and Adaptation Strategies for Global Change*. http://doi.org/r46. Accessed 15 Dec 2014.

Rao, M., Johnson, A., Spence, K., Sypasong, A., Bynum, N., Sterling, E., & Praxaysombath, B. (2014). Building capacity for protected area management in Lao PDR. *Environmental Management, 53*, 715–727. https://doi.org/10.1007/s00267-014-0235-9

Rauken, T., Mysdske, P. K., & Winsvold, M. (2015). Mainstreaming climate change adaptation at the local level. *Local Environment, 20*, 408–423. https://doi.org/10.1080/13549839.2014.880412

Reisinger, A., Meinshausen, M., & Manning, M. R. (2011). Future changes in global warming potentials under representative concentration pathways. *Environmental Research Letters, 6*, 024020.

Robson, C. (2002). *Real world research, a resource for social scientists and practitioner researchers*. Blackwell Publishers.

Saunders, M., Lewis, P., & Thornhill, A. (2000). *Research methods for business* (2nd ed.). Pearson Education Limited.

Scoones, I. (2009). Livelihoods perspectives and rural development. *The Journal of Peasant Studies, 36*(1), 171–196.

Schakel, A.H., Hooghe, L. & Marks, G. (2015) *Multilevel Governance and the state*. In: Leibfried, S. et al. (eds), The Oxford Handbook of Transformations of the State. Oxford Academic, 10 Sept. 2015.

Sem, G. (2007). *Vulnerability and adaptation to climate change in Small Island Developing States*. United Nations Development Programme.

Sen, A. A. S. (1997) *Concepts of Human Development and Poverty: A Multidimensional Perspective*. In: Poverty and Human Development: Human Development Papers 1997. New York: United Nations Development Programme. 1–20.

Sen, A. (1999). *Development as freedom* (1st ed.). New York: Oxford University Press

Setó-Pamies, D. & Papaoikonomou, E. (2016) A multi-level perspective for the integration of ethics, corporate social responsibility and sustainability (ECSRS) in management education. *Journal of Business Ethics, 136*, 523–538.

Shakya, C., Cooke, K., Gupta, N., Bull, Z., & Greene, S. (2018, October). *Building institutional capacity for enhancing resilience to climate change: An operational framework and insights from practice: Action on Climate Today (ACT)*. Oxford Policy Management & IIED.

Sharma, A. (2018). *Pocket Guide To capacity-building: For Climate change*. First published May 2017: International Climate Initiative (IKI): Updated March 2018.

Sherman, M., & Ford, J. (2013). Stakeholder engagement in adaptation interventions: An evaluation of projects in developing nations. *Climate Policy*. http://doi.org/r45. Accessed 15 Dec 2014.

Simmons, A. (2018). Impact of climate change on youth in small island communities: The case of St. Vincent and the Grenadines: This thesis is submitted to De Montfort University in partial fulfilment of the requirements for the award of Doctor of Philosophy: School of Engineering and Sustainable Development De Montfort University.

Simmons, A. (2021). *Impact of climate change on youth in Small Island Developing Communities* (Sustainable development series). Springer Nature.

Smit, B., & Wandel, J. (2006). Adaptation, adaptive capacity, and vulnerability. *Global Environmental Change, 16*, 282–292. https://doi.org/10.1016/j.gloenvcha.2006.03.008

Stake, R. E. (1995). *The art of case study*. Sage.

Stiller, S., & Meijerink, S. (2016). Leadership within regional climate change adaptation networks: The case of climate adaptation officers in Northern Hesse, Germany. *Environmental Change: Cross Mark, 16*, 1543–1555.

Tourney, D. (2019). Follow the leader? Conceptualising the relationship between leaders and followers in polycentric climate governance. *Environmental Politics. 28*(1),167–186.

Tschakert, P., & Dietrich, K. A. (2010). Anticipatory learning for climate change adaptation and resilience. *Ecology and Society*, 15(2), 11.

Turnbull, M., Sterrett, C. L., & Hilleboe, A. (2013). *Toward resilience: A guide to disaster risk reduction and climate change adaptation*. Practical Action Publishing Ltd.

Udie, J., Bhattacharyya, S., & Ozawa-Meida, L. (2018). A conceptual framework for vulnerability assessment of climate change impact on critical oil and gas infrastructure in The Niger Delta. *Climate, 6*(1), 11.

Uhl-Bien, M., Marion, R., & McKelvey, B. (2007). Complexity leadership theory; shifting leadership from the industrial age to the knowledge era. *The Leadership Quarterly, 18*, 298–318. https://doi.org/10.1016/j.leaqua.2007.04.002

Uittenbroek, C. J., Janssen-Jansen, L. B., & Runhaar, H. A. (2013). Mainstreaming climate adaptation into urban planning: Overcoming barriers, seizing opportunities and evaluating the results in two Dutch case studies. *Regional Environmental Change, 13*, 399–411. https://doi.org/10.1007/s10113-012-0348-8555. https://doi.org/10.1007/s10113-015-0886-y

UNDP. (2004). *Adaptation policy frameworks for climate change: Developing strategies, policies, and measures*. Cambridge University Press.

UNDP. (2009). *Capacity development: A UNDP primer*. United Nations Development Programme. Available at: https://www.undp.org/publications/capacity-development-undp-primer

UNDP. (2010). *Designing climate change adaptation initiatives: A UNDP toolkit for practitioners*. United Nations Development Programme.

UNDP. (2015). *Human development report*. Work for Human Development, United National Development Programme (UNDP).

UNFCCC (2017). *National Adaptation Plans*. Available online: https://unfccc.int/topics/adaptation-and-resilience/workstreams/national-adaptation-plans

United Nations (UN). (2015). Resolution adopted by the General Assembly on 25 September 2015. A/RES/70/1 Transforming our world: The 2030 Agenda for Sustainable Development.

United Nations Environment Programmes. (2013). *Identifying emerging issues from the perspective of Small Island Developing States (SIDS)*. Expert Group Meeting Co-hosted by United Nations Environment Program (UNEP) and the United Nations Department of Economic and Social Affairs (DESA), 14–16 May 2013, Cambridge, United Kingdom.

United Nations Framework Convention for Climate Change (UNFCCC). (1992). Article 6 of the Convention Draft conclusions proposed by the Chair Recommendation of the Subsidiary Body for Doha work programme on Article 6 of the Convention, (December) (pp. 1–11). Retrieved from http://unfccc.int/cooperation_support/education_outreach/overview/items/8950.php

United Nations Framework Convention for Climate Change (UNFCCC). (2001). Report of the Conference of the Parties on its seventh session, held at Marrakesh from 29 October to 10 November 2001. Addendum. Part two: Action taken by the Conference of the Parties. Volume I, 11, 111 & Iv.

USAID (2012) *USAID 2012 Sustainability Plan. Executive Order 13514*. Available at: https://2017-2020.usaid.gov/sites/default/files/documents/1865/USAID2012AgencySustainabilityPlan.pdf

Virji, H., Padgham, J., & Seipt, C. (2012). Capacity-building to support knowledge systems for resilient development—Approaches, actions, and needs. *Current Opinion in Environmental Sustainability, 4*(1), 115–121. https://doi.org/10.1016/j.cosust.2012.01.005

Wamsler, C., Luederitz, C., & Brink, E. (2014). Local levers for change: Mainstreaming ecosystem-based adaptation into municipal planning to foster sustainability transitions. *Global Environmental Change, 29*, 189–201. https://doi.org/10.1016/j.gloenvcha.2014.09.008

Whittle, S., Colgan, A., & Rafferty, M. (2012). *Capacity-building: What the literature tells us*. The Centre for Effective Services.

Williams, C., Fenton, A., & Huq, S. (2015). Knowledge and adaptive capacity. *Nature Climate Change, 5*, 82–83. https://doi.org/10.1038/nclimate2476

Williams, D. S., Sérgio, R., Oocheetsing, S., & Louis, C. (2020). Identifying local governance capacity needs for implementing climate change adaptation in Mauritius.; Taylor and Francis. *Climate Policy, 20*(5), 548–562. https://doi.org/10.1080/14693062.2020.1745743

Wise, R. M., Fazey, I., Stafford Smith, M., Park, S. E., Eakin, H. C., Archer Van Garderen, E. R. M., & Campbell, B. (2014). Reconceptualising adaptation to climate change as part of pathways of change and response. *Global Environmental Change, 28*, 325–336. https://doi.org/10.1016/j.gloenvcha.2013.12.002

World Bank. (2005). *Capacity-building in Africa: An independent evaluation*. World Bank.

World Bank. (2009). Comprehensive development framework, a discussion draft. Available at: www.worldbank.org/cdf/

Yin, R. K. (2014). *Case study research: Design and methods*. Sage.

Yin, R. K., & Campbell, D. T. (1989). *Case study research: Design and methods*. University of California/Sage.

3 The Caribbean Perspective: Building the Capacity of National Institutions as a Strategy for Enhancing the Resilience of People and Communities to the Impact of Climate Change in St. Vincent and the Grenadines

Abstract

The chapter presents the research methods used to collect data for the case studies. This is followed by a case study on the work of these community organizations in Enhams, St. Vincent and the Grenadines, working with communities to build their resilience to adapt to the impact of climate change and other development calamities. The chapter maps the JEMS history of environmental sustainability throughout four and a half decades. The organization was awarded the National Youth Award in 1985 and 1986 and the Caribbean Conservation Award in 1993 for its innovative community conservation approach to protecting the natural environment by utilizing cultural art forms, including popular music and poetry, to empower and raise awareness of residents of the southeast of mainland St. Vincent and the Grenadines. The founder and former president of the organization Andrew Simmons won the UN Global 500 Award in 1990 and the Goldman Environmental Prize in 1994 for leadership in preserving and conserving the global environment through JEMS.

The chapter then presents a case study on the work of the Government of St. Vincent and the Grenadines Community Development Division, the Ministry of Community Services, during the period 1990–2000. The institution initially utilized top-down development approaches to develop the community development programmes in St. Vincent and the Grenadines. The case highlights the shift to a hybrid participatory "aided self-help" basic needs approach to implementing the community development programmes after changes in leadership and programmes of the organization. The case study tracks the strategic methods implemented by the division to mobilize local communities to participate in its programmes to address skills employment shortages, enhance basic knowledge and skills on development, build community infrastructure and support sustainable livelihoods. The chapter ends with a summary of the main findings from the case studies.

Keywords

Sustainable livelihood · Community development programme · Employment · Community infrastructure · Top-down approach, Bottom-up approach, Participatory approach, Basic needs, Poverty, Sustainable development.

The first part of the chapter presents information on the research methods used to collect data for the case studies. This is followed by a case study on the work of the JEMS Progressive Community

A. Simmons, A.-A. Simmons, *Building Institutions for Resilience*, Sustainable Development Goals Series, https://doi.org/10.1007/978-3-031-28881-4_3

Organization in Enhams, St. Vincent and the Grenadines, working with communities to build their resilience to adapt to the impact of climate change and other development calamities. The chapter charted the JEMS history of environmental sustainability over four and a half decades. The organization was awarded the National Youth Award in 1985 and 1986 and the Caribbean Conservation Award in 1993 for its innovative community conservation approach to protecting the natural environment by utilizing cultural art forms, including popular music and poetry, to empower and raise awareness of residents of the southeast of mainland St. Vincent and the Grenadines. The founder and former president of the organization Andrew Simmons won the UN Global 500 Award in 1990 and the Goldman Environmental Prize in 1994 for leadership in preserving and conserving the global environment through JEMS.

The third part of the chapter presented a case study on the work of the Government of St. Vincent and the Grenadines Community Development Division, the Ministry of Community Services, during the period 1990–2000. The institution initially utilized top-down development approaches to develop the community development programmes in St. Vincent and the Grenadines. The case highlighted the shift to a hybrid participatory "aided self-help" basic needs approach to the implementation of the community development programmes after changes in leadership and programmes of the organization. The case study tracked the strategic approaches implemented by the division to mobilize local communities to participate in its programmes to address skills employment shortages, enhance basic knowledge and skills on development, build community infrastructure and support sustainable livelihoods. The chapter ends with a summary of the main findings from the case studies.

3.1 Research Methods to Collect Data for Case Studies

Collating the data for these case studies started with a desk review of all documents produced over the past 40 years by JEMS, the Community Development Division, the Government of St. Vincent and the Grenadines (Robson, 2002; Angelsen et al., 2011). The researchers analysed these documents to identify the strengths and weaknesses of the organizations, the challenges encountered by the organizations when implementing capacity-building programmes, the gaps in methodologies and strategies utilized by the organizations and the effectiveness of these strategies to mobilize the population to participate in their development programmes. Assessments were also conducted on the effectiveness and relevance of the programmes to meet the development needs of the communities assisted by these organizations. The information extracted from these documents was integrated into the case studies.

Interviews were conducted with a broad group of stakeholders, including former staff, members and project beneficiaries of both participating organizations. The performance of the organizations was assessed by identifying the strengths and weaknesses of the organizations, the strategies and projects that worked well and those that did not work well, the lesson learned and best practices and recommendations to enhance the overall quality of performance of both institutions. Focus group discussions (FGD) were conducted with members of both institutions and beneficiaries from the community to assess the strategies implemented to enhance the quality of the capacity-building programmes implemented.

The multi-research methodologies were implemented to collect data. These methodologies, including desk reviews, interviews, questionnaires and focus group discussions (FGDs), provide two significant advantages to enhance the quality of output of the research process:

- It provides the research process with the confidence necessary to pursue the research.
- It enables triangulation to occur, i.e. using different data collection systems within a study to ascertain that the data given is the correct information (Saunders et al., 2000; Robson, 2002).

3.2 Population Sampling

A good sampling strategy aims to ensure that appropriate data is collected at each level relevant to the nature of the analysis to be implemented by the research study team (Angelsen et al., 2011). This ensures that data from the different levels are brought to bear in understanding the social, political and physical environment in which stakeholders operate. Generally, the sampling is implemented at several stages of the study (research process) (Robson, 2002; Byrne, 2002 in Angelsen et al., 2011).

A similar sampling process was instituted to select the beneficiaries participating in the data collection process. After zoning the country, the names of all the partner institutions and communities are compiled and chosen randomly. It is important to note that depending on the population size to be studied, the sample size can be as small as 10% of the population that benefitted from the project. It is also important to note that this is not normal, especially regarding the restrictions and risks associated with the coronavirus pandemic.

3.3 Data Collection

As stated, the multi-methodology approach was used to collect data to prepare all the case studies for the book. The details of how these were carried out are explained below.

(a) *Desk Review*

The desk review included a review of all the relevant reports and documents on the work of JEMS and the Community Development Division (CDD) by collecting, organizing and synthesizing available information to present an understanding of the role the organizations played in enhancing the capacity of the people within the organizations and communities to contribute to the protection of the environment and the achievement of sustainable livelihood outcomes. Reports on the performance of the organization were analysed. The analysis was conducted by examining the strengths and weaknesses of the organization, including the relevant organizational arrangements and partnerships.

(b) *Interviews*

Semi-structured interviews were conducted with members of staff, former members of the organizations and beneficiaries of the organization and the officials from appropriate government agencies to solicit their views and concerns on the role of JEMS and the CDD as a development organization and to tease out relevant information that was not documented. This interview format allowed authors to establish personal contact with officials of the organization, the appropriate governmental agencies and partner institutions/agencies, thus facilitating the building of trust and cooperation with the officials (Robson, 2002; Angelsen et al., 2011).

During the research, consultative interviews/discussions were conducted among officials and representatives from partner institutions. Questions were asked on the (a) strengths and relevance and design of the programmes/project, (b) weaknesses of these development projects, (c) problems regarding the project implementation and operational process, (d) lessons learned, (e) best practices and (f) recommendations for the future.

(c) *Questionnaires*

Due to the complexity and challenges caused by the outbreak of the coronavirus pandemic and the implementation of social distancing guidelines globally, this situation posed unique difficulties in collecting data from the CDD and JEMS members, former staff and beneficiaries and representatives of partner institutions. After discussion with the necessary stakeholders, questionnaires were selected as one of the preferred data collection tools. To reduce the incidence of data contamination, we used personal emails and requested participants to return the emails to us as soon as they completed the forms.

(d) *Observation*

Although the observation approach can be prone to numerous ethical considerations, I view it as a valuable research tool and conduct it regularly when designing and implementing projects. For example, I ran a series of observations in St. Vincent and the Grenadines from October to December 2021 to evaluate the delivery of the work of JEMS and CDD by visiting project sites in communities to meet with beneficiaries and observe whether these projects were appropriate and still functioning.

Beneficiaries were informed in advance of our visits and the nature of our visits. However, this prior information possessed the potential to lead to non-participation or altering behaviour in such a way as to present invalid information, a phenomenon described as the Hawthorne effect (Fox et al., 2008). The beneficiaries of the projects were very supportive of the research process.

(e) *Case Studies*

Participatory research is a family of approaches, behaviours and methods to enable people to do their appraisal, analysis, planning, monitoring and evaluation (Chambers, 2002). The methods identified for collecting data are relevant to the research's aims and objectives, the research questions and the target population (Angelsen et al., 2011).

We compiled a small number of case studies as part of a comparative study. As advocated by Yin (2014), we used the case study methodologies, which included (a) using multiple sources of evidence, such as archival records, interviews and direct observations, (b) establishing the chain of evidence, (c) including key informants in reviewing draft study report, (d) carrying out pattern matching or explanation building or time series analysis, (e) applying replication logic in multiple case studies, (f) using study protocol and (g) developing case study database tools.

3.4 Case Study 1: JEMS Progressive Community Organization – Participatory Integrated Development Model for Bottom-Up Participation of Communities in Building Capacity for Sustainable Development

3.4.1 Historical Context

Before 1972, about 95% of the people in the 15 villages in the southeast of mainland St. Vincent and the Grenadines were employed on 4 agricultural estates farming arrowroots, cotton, groundnuts and other root tubers. The villagers also worked in manufacturing starch from the rhizomes of the arrowroot, which were exported to the USA and the UK to coat the papers of large computers for food supplements and the raw material for various products in the cosmetics and pharmaceutical industry. The oil crisis in 1972 and the increased fuel cost resulted in the closure of these agricultural estates and the arrowroot manufacturing facilities. Unemployment soared from a mere 5% in the villages to as high as 90%, plunging the communities into mass unemployment and poverty (Ritten & LaGra, 1991; https://www.goldmanprize.org/recipient/andrew-simmons/).

One of the oldest forest reserves in the world, King's Hill Forest Reserve, is located on this part of the island. It was established on 52 acres of land in 1791 when "the Caribbean was in the middle of the worst drought known in its history, caused by the longest and severest El Nino event in written history…one consequence of this event was the passing of a pioneering piece of conservation legislation. This was the Kings Hill Forest Act" (Grove, 2000, pp. 1–2).

The oil crisis impacted the lives of villagers; to ensure their survival, they began cutting down trees in King's Hill Forest Reserve and along the riverbanks for fuel wood and poaching for food. This situation was compounded by the destruction of wildlife in the Milligan Cay Bird Sanctuary, an area designated as a reserve for migratory birds and wildlife. Residents began patrolling the beaches of Brighton, Diamond and Stubbs in search of endangered turtles and other species for food (Leach, 1996).

The river flowing through the village of Enhams dried up in 1977 and prompted a group of young people to begin discussions with other villagers' solutions to reduce the high level of poverty and halt the environmental destruction in the community. One year later, 52 young people formed an organization called the JEMS Progressive Community Organization. When we started going into the community, telling people to stop cutting down the trees, I learned a harsh lesson when confronted by one woman who emphatically stated: "How do you expect me to get food? I can't buy natural gas. How do you expect me to get food to feed my children and to send them to school?"

After this enlightenment, we decided to change our strategy to one where we work with villagers to save King's Hill Forest Reserve while also providing skills training to locals of the area to secure sustainable livelihoods. One of the significant problems in these communities was illiteracy. It was estimated that over 70% of adults were unable to read, complete job application forms or write a letter to their parliamentary representative to complain about the deforestation problems in the community. A literacy programme was instituted, providing classes to over 300 villagers 4 evenings per week. Primary and secondary school teachers, students, skilled masons and carpenters were recruited and trained as adult learning instructors (Leach, 1996). The Ministries of Education and Community Development were engaged by JEMS, and schools and community centres were provided to the organization to implement these community programmes.

3.4.2 Overview of St. Vincent and the Grenadines: Geographic and Communities Supported by JEMS

The island of St. Vincent is approximately 35 km long and 15 km wide. Volcanic hills extend their steep slopes from the land with narrow shelves that can be found on the leeward side of the island. A slightly wider shelf is located on the windward side of the island. These areas are prone to landslides and flooding during hurricane and storm events. St. Vincent has over 100 km of coastal area with predominantly black volcanic sandy beaches on the mainland, while in the Grenadines, the beaches are mainly white coralline sand. Most of the beaches on mainland St. Vincent have disappeared to feed the construction industry's expansion and erosion due to storm surges and sea level rise (CARIBSAVE, 2012).

St. Vincent and the Grenadines is a member of a category of countries called small island developing states (SIDS). SIDS, comprising 58 countries, form a distinct group of developing countries across 3 geographic regions spanning the ocean regions of the Pacific, Indian and Atlantic, as well as the Mediterranean and Caribbean Seas, and stand to be among the first and foremost states diversely affected by the impact of climate change (Lazarus, 2012; IPCC, 2014a, b, c; Hernandez-Delgado, 2015). Sea level rises, changing precipitation and rainfall regimes, changes in tropical storms and cyclone frequency or intensity, increasing air and sea surface temperatures, changes in atmospheric chemistry and melting of mountain glaciers are some of the phenomena and disruptions related to climate change (Gomez, 2013; IPCC, 2014a, b, c; Simmons, 2021).

The demographic served by JEMS had shifted considerably from 1978 when the organization began its operation. Back then, the 15 villages were comprised mainly of unemployed youths and women with high levels of functional illiteracy. With the closure of the major agricultural estates, residents had to seek alternative sources of survival. The result was squatting, deforestation, beach erosion and the near destruction of King's Hill Forest Reserve, the oldest forest reserve in the western hemisphere, established in 1791 (Grove, 2000). Today, that demographic has shifted to far more educated and skills-based constituents, coupled with high unemployment, poverty, crime, drugs and HIV/AIDS levels. It is estimated that over 20,000 thousand people live in these communities, approximately one-fifth of the population (Simmons, 2018, 2021; Rittgens & LaGra, 1991).

JEMS' work mainly centred in the 15 villages southeast of mainland St. Vincent and the Grenadines. During the review period, the JEMS provided technical and development support and capacity-building, including training, coaching, mentoring support and networking to over 40 villages, NGOs and civil society organizations across St. Vincent and the Grenadines (Simmons, 2018, 2021).

3.4.3 Overview of the JEMS Organization Project

Established in 1978, the JEMS Progressive Community Organization is a grassroots community development and environmental management organization located in St. Vincent and the Grenadines in the Eastern Caribbean. The original name JEMS derives from its founding rural communities southeast of mainland St. Vincent, namely, Junction, Enhams, McCarthy and surrounding areas.

JEMS' geographic location is home to several national environmental sites and artefacts on mainland St. Vincent. These include King's Hill Forest Reserve, turtle-nesting beaches, the 3000-year-old archaeological Amerindian sites currently being destroyed by sand mining and the Milligan Cay Bird Sanctuary, home to thousands of migratory birds. This bird sanctuary is located directly in the path of Argyle International Airport and may have consequences for planes flying into St. Vincent. There are also mangrove swamps on Brighton and Salt Pond beaches facing extinction (https://www.goldmanprize.org/recipient/andrew-simmons/: Rottgens & La Gra, 1991).

Throughout its 45 years as an organization, JEMS has expanded from programming in 3 established villages to include working directly with over 20,000 people within 15 area communities in the southeast and providing support to over 40 villages, community-based organizations (CBOs), youth environment school clubs (YES) and other environment and development organizations in St. Vincent and the Grenadines. With the expansion of its programmes to communities throughout St. Vincent and the Grenadines and the implementation of cooperative projects with neighbouring islands and nations in the Caribbean, JEMS has grown from a grassroots community-based organization to one of national and regional influence. Through support from the United Nations Environment Programme (UNEP) and the Caribbean Conservation Association (CCA), JEMS established the Caribbean Youth Environment Network (CYEN) to replicate its bottom-up participative integrative development model, engaging youth from across the Caribbean region to tackle environmental and development challenges (OECD, 2013; Matland, 1995; Stochowiak et al., 2016).

JEMS is one of the few registered NGOs in St. Vincent and the Grenadines that remain active and have its own headquarters. With a legacy of longevity, its success is based primarily on a bottom-up "self-help", people-first philosophy, which has yielded results according to the organization's profile on national, regional and international levels (Leach, 1996; Rittgens & La Gra, 1991). The JEMS brand has since expanded to include the following entities: JEMS Progressive Community Organization, JEMS Environmental Services Organization, JEMS Institute for Popular Education, JEMS Pre-school, JEMS Theatre (New York, USA), JEMS for Hope (New York, USA) and JEMS Youth and Community Enterprise Development Initiative (JEMS YACEDI), Tottenham, London.

3.4.3.1 Collaboration and Partnership with Development Partners

Besides the knowledge and skills of its membership and its wide range of innovative participatory strategies/methodologies for mobilizing and enhancing the participation of people in the environment and sustainable development activities, JEMS did not have the necessary finances to implement its development programmes at national and regional levels. The organization effectively formed partnerships with various development agencies at national, regional and international levels to ensure that resources are mobilized and channelled to enhance the people's lives in the communities it serves.

Over the years, the organization has collaborated with governmental agencies, regional and international development partners, private sector agencies, IGOs and bi–/multilateral agencies to implement its programmes to implement environmental and sustainable development projects. At the national level, it collaborated with the Ministry of Health and Environment; Ministry of Agriculture, Forestry and Fisheries; Ministry of Culture and Tourism; Ministry of Youth Affairs; Community Development Division in the Ministry of Social Development; South Eastern Development Inc.; National Association of Development Organizations (NADO); National Association of Mass Education (NAME); National Youth Council (NYC); and Lions Club South to implement a range of environmental and sustainable development projects at community and national levels.

Regionally, JEMS has collaborated with the University of the West Indies (UWI), the Organisation of Eastern Caribbean States (OECS), the National Resource Management Unit, the Caribbean Youth Environment Network (CYEN), the Caribbean Natural Resources Institute (CANARI), the Caribbean Regional Council for Adult Education (CRCAE), the Caribbean Conference of Churches and the Caribbean Conservation Association (CCA) to implement environmental education and sustainable capacity-building programmes in St. Vincent and across the Caribbean. For 2 years, the Caribbean Conference of Churches was instrumental in funding the JEMS Adult Education and Literacy programmes in St. Vincent and the Grenadines. The programme ran for over a decade and graduated over 3000 adult learners in literacy and skills for employment and sustainable livelihoods certification. The JEMS conservation education and capacity development initiative was funded by the University of the West Indies and certified over 100 community leaders in leadership for sustainable development from St. Vincent and the Grenadines.

At the international level, JEMS collaborated with the Ocean Conservancy (USA), United Nations Environment Programme (UNEP), Goldman Environmental Prize, Jewish Community Federation, UNDP GEF Small Grants Programme and United Nations Development Programme (UNDP) to enhance the capacity of leaders to conduct research on the pollution of the ocean and seas and manage environmental conservation education programmes across the Caribbean. It mobilized financial support from the Goldman Environmental Prize and the Jewish Community Federation to implement the National Climate Change Campaign project in St. Vincent and the Grenadines. It mobilized over 45,000 people in St. Vincent and the Grenadines to implement climate action activities in communities across St. Vincent and the Grenadines.

An extensive amount of environmental work was implemented with support from the Global Environment Facility Small Grants Programme (GEF SGP) to increase communities' understanding of the GEF thematic areas through focus group discussion of GEF SGP structures and methodologies and equip 25 community leaders with skills in project writing, project implementation and evaluation. It also established a Project Appraisal Committee to prioritize pre-screening all potential GEF SGP in St. Vincent and the Grenadines before sending them to the Barbados-based Project Selection Committee for funding. The Project Appraisal Committee was approved by the Government of St. Vincent and the Grenadines and the UNDP to support the screening, approval and implementation of environmental projects in St. Vincent and the Grenadines (https://sgp.undp.org/special-itemid; https://www.goldmanprize.org/recipient/andrew-simmons/).

3.4.3.2 Aim and Objectives of the Projects/Programmes

Over the past 40 years, JEMS has undergone a process of revolutionary change. It has evolved from an organization bent on protecting the environment and livelihoods of the people and their communities southeast of mainland St. Vincent and the Grenadines to one leading the charge on implementing climate change and sustainable development projects across the Caribbean. Now the organization is facing a period of dormancy. The only portion of the active organization is its preschool and daycare facility.

The mission statement of JEMS states that:

> JEMS is a national environmental network in St. Vincent and the Grenadines with an emphasis on enhancing the institutional capacity of CBOs to manage community resources and develop life-sustaining practices.

Its objectives are:

- Foster a spirit of understanding and unity among member organizations and the community working on the environment and sustainable development.
- Facilitate "self-help" projects that respond to the needs of people and will assist in improving the quality of life for community members.
- Work to develop qualities of leadership in people within development organizations and communities across St. Vincent and the Grenadines.
- Collaborate with other local, regional and international organizations and institutions involved in environmental and development issues to reduce poverty through enhancing sustainable livelihoods, decrease the impacts of climate change by building people's resilience to adapt to climate change's impact and overall develop sustainable communities.

3.4.4 Strategies and Methodologies Used to Mobilize and Build the Capacity of Communities

JEMS utilizes popular cultural activities to mobilize young people, women and their communities to participate in environmental and sustainable development actions through its cultural wake strategy (Leach, 1996; Simmons, 2018, 2021). The cultural wakes combine music (steel pan, drumming), drama and discussion, which not only entertain but provide opportunities for participants to identify major issues affecting their community and devise strategies to respond to the development challenges they encounter. This cultural wake strategy was used to mobilize resources and skilled workers to provide the communities with important basic needs projects such as pipe-borne water, climate risk and sustainable housing (rebuilding of housing stocks after they were destroyed by hurricanes and storms), community and coastal clean-up programmes, basic education/literacy and sustainable health and sanitation projects. The cultural wakes were also utilized to mobilize the community to participate in environmental and sustainable development projects. In the 2000s, it mobilized communities to participate in environmental conservation and climate change campaigns, poverty alleviation and HIV/AIDS prevention campaigns (Leach, 1996; Chambers, 2002).

The participatory community mobilization initiative implemented by JEMS has a series of development stages. The first stage of the process consisted of conducting research into the community to understand the nature and extent of the needs and issues faced by the community. This stage consisted of house-to-house consultations, participatory community mapping and focus group discussions with residents on collecting data to determine the extent and severity of the community's development challenges.

The second stage of the process consisted of organizing cultural activities in the community to raise residents' awareness about the problem encountered and explore what can be done to resolve these problems. The community would meet for approximately a week to practice cultural activities, including drumming, calypsos, skits and poetry, to perform at the cultural wake at the village square on a moonlight night. The entire community will attend the cultural wakes. They will invite artists from other communities to participate in the activities and present various renditions/compositions on the issues facing the communities to the community. At the end of the performance, a community animator would facilitate discussions on the strategies presented by and for the community to resolve the development problems facing the communities (Leach, 1996; Simmons, 2018, 2021).

In the third stage, the community will select a set of representatives who will meet to plan programmes to resolve the problems. Here, other representatives from each appropriate governmental agency would be invited as resource persons to participate in the community planning process. In a situation where there were no community centres and/or public facilities in the village, meetings would be held in a village run-shop and/or at the home of one of the village residents to collectively work out the solutions to resolve the problems. Where the solution to determine the development challenges affecting the community may require outside material and resources injection, a senior member of JEMS will be selected to work with the community to prepare project proposals for submission to governmental agencies – such as the Community Development Division, Parks and Protected Areas, Ministry of Agriculture. External funding agencies such as UNDP GEF SGP programmes operated by the US and Canadian Embassies in Barbados will be targeted to provide financial and personal support to the project (https://sgp.undp.org/special-itemid).

Stage 4 of the process includes the project implementation when the funding is secured. The community would provide labour and food matched with the material and technical assistance provided by the funding agency. Monitoring and evaluation are entrenched as essential tools throughout the project implementation process. Meetings would be organized in communities on a fortnightly basis to reflect on the achievement and pitfalls of the previous stage and agreement on planned activities for the next stage. At all stages during the capacity-building process, the community leaders were trained on all aspects of project management.

During the period under review, several communities were without public roads, pipe-borne water and sanitary facilities such as toilets and public baths. JEMS exploited the cultural wake models to mobilize communities to implement pipe-borne water projects in the villages of McCarthy, Carapan, Choppins, Argyle and Lower Stubbs. It constructed public baths and sanitary garbage facilities in Lower Stubbs and Enhams. The provision of pipe-borne water to villages was a watershed initiative for JEMS since, previously, villagers would have to collect water from the rivers or from other villages located miles away to meet their daily domestic needs. Children and youth involved in the pipe-borne water and fuel wood collection were robbed of the time they could have spent acquiring an education. JEMS also worked with the Argyle RC primary school staff to deliver pipe-borne water to the school. The community of Argyle and Peruvian Vale was overwhelmed by the project. The JEMS projects enhanced the health, sanitation standards and well-being of people in the villages. The cultural wake models were transferred to other communities across St. Vincent and the Grenadines through the work of the Government's Community Development Division, the National Youth Council of St. Vincent and the Grenadines and other affiliates of JEMS (Rittgens & La Gra, 1991; Leach, 1996).

3.4.5 Challenges and Risks Faced by the Organization and Communities

Notwithstanding its outstanding record of success and achievements, JEMS has had to struggle with several challenges that may have caused other organizations to stop functioning. Among the challenges was a meeting place to convene meetings in its formative years. JEMs can boast of being one of the few NGOs in St. Vincent and the Grenadines with its own building due to collaborative efforts involving several vital partners.

The migration of critical members has also impacted the organization considerably. Many of its original members migrated overseas, mainly to the USA, Canada and the UK, to seek sustainable livelihoods. Many trained builders (carpenters, masons, etc.) who graduated from the JEMS Institute of Popular Education (JIPE) travelled to other Caribbean countries in search of employment. The implication of this phenomenon has resulted in JEMS constantly mobilizing new members and putting measures in place to sustain its membership within the communities. The organization can boast of having over 1500 volunteers dispersed throughout the communities in southeast St. Vincent. A positive impact of migration on the organization is that these members have established similar community-based organizations in inner cities in the USA, Canada and the UK to support the organization's development work in St. Vincent and the Grenadines.

The organization has no full-time staff and operates with 100% volunteers. JEMS work flourished due to its robust, effective and vibrant leadership. This has resulted in the organization entering a dormancy stage after I migrated to work with an international agency. When I returned in 2006 to St. Vincent, this prompted the resuscitation of JEMS and the injection of new impetus, which has seen the emergence of the organization as a leading voice and advocate for several issues, including the environment, HIV/AIDS, climate change and sustainable livelihoods. Although the organizations implemented numerous leadership capacity-building training for the new members, keeping these young people in the country is challenging due to the high unemployment among youth in St. Vincent and the Grenadines, estimated at over 32%. The older leaders are now occupied with caring for grandchildren and careers. Another issue facing the organization is that doing community work is no longer attractive to the current crop of young people. The policies implemented by the current government do not support community volunteerism. Residents complained that government agencies would employ the ruling party's supporters to implement basic needs projects in the community and ignore those who would usually participate in volunteer work.

Understandably, financial resource, or lack thereof, has been one of the significant challenges confronting the organization. Its survival has hinged more on the generosity and voluntarism of its membership and the community it served. Much of this has been cushioned by community donations and material support from the grassroots level. While receiving some donor support for project activities, JEMS failure to attract and secure a sound financial base continues to be a significant challenge, if not the major constraint, particularly given the harsh global economic climate.

As an advocate of national environmental issues, JEMS has been strongly criticized because of its stance on government destructive policies and projects such as the Diamond Quarry, the construction of the solid waste facility at Diamond, the Cross-Country Road Project and the Argyle International Airport. JEMS motivated the communities to rally and picketed government departments, halting the mining project in Diamond. Three years after this seeming victory, work resumed at the quarry, and the challenge to this unsustainable development practice prevailed. This marked the first significant variance with the government on issues of this nature. More recently, JEMS disagreed with the government concerning the proposed establishment of a cross-country road and its failure to conduct an environmental impact assessment (EIA) in respect of the Argyle International Airport. After many talks, the government conceded and implemented EIAs for both projects.

3.4.6 Strategies Implemented to Address the Above Challenges/Risks

Although JEMS has been effective in devising strategies to address the challenges and risks facing the organization, in recent years, these contentious issues have impacted JEMS' operations and effectiveness as a national NGO. JEMS' position on the Cross-Country Road and the Argyle International Airport projects has resulted in ongoing hostility with the current government, which has affected JEMS' work and its relationship with certain branches of the government. As a result, some of the JEMS' work and agenda have been somewhat "sabotaged" from time to time, side-lined by perceptions relative to partisan politics.

However, despite these odds, JEMS' work and impact have persevered. While this has and will continue to pose a challenge in small developing island states such as St. Vincent and the Grenadines, the respect the organization has earned at the community, national and international levels has always triumphed over such difficulties. Over the past few years, JEMS has been working continuously to enhance its relationship with the central government while simultaneously advocating for the rights of ordinary people and their communities.

3.4.6.1 Capacity Development Programmes Implemented by JEMS

JEMS has developed and implemented programmes that evolved from the reality villagers were experiencing and contributed to addressing the community's basic needs. These strategic programmes are:

3.4.6.2 Adult Education and Literacy

JEMS created the JEMS Institute of Popular Education (JIPE) in 1985 to implement a decade-long adult literacy, skills development and education programme which provides opportunities for over 3000 persons from the villages to develop skills in numeracy, literacy and sustainable livelihoods skills which led to employment and improvement in health, education and human well-being. These classes were held four evenings per week. The curriculum included academic subjects such as English and Mathematics. It provided skills training in carpentry, masonry, sewing and design, sustainable agriculture and home economics. It also provided classes in civic responsibilities, including personal skills development, arts and culture and community and national development. JIPE produced carpenters, masons, auto mechanics and electricians to work in 20 communities in the southeast of mainland St. Vincent and the Grenadines (Leach, 1996).

Through its education programme, the organization raised the level of literacy and civic responsibility, thus creating an environment where villagers felt more empowered to fully participate in voting for the government of their choice and/or writing to their parliamentary representatives to demand quality services. The initial stages of the project were funded by the Caribbean Conference of Churches. The construction skills training aspect of the project was supported by the Government Community Development Division through support from DfID and USAID. Finding sustainable employment was a major challenge facing the graduates of JIPE until local businessman O.T. Mayers, proprietor of Finishing and Furnishing, started employing them in his furniture factory at Diamond and on various construction sites.

I was so excited that I would return to St. Vincent and the Grenadines every year to train his furniture factory staff on business practices, safety standards in furniture manufacturing and sustainable livelihood practices.

Therefore, to ensure sustainable livelihoods for graduates, it is necessary for training providers such as JEMS and governmental institutions to forge a partnership with private sector companies to contribute to the development of curriculum, provisions of equipment and internship opportunities for learners and graduates. This partnership would ensure that the skills and competencies provided by these institutions are relevant and appropriate to the economy's needs (Simmons, 2018, 2021; ILO, 2010, Poschen, 2015; UNDP, 2015).

3.4.6.3 Community Cultural Wake Initiative

JEMS used "Cultural Moonlight Wake" to sensitize and mobilize residents to participate in its environment and community development programmes. Cultural wakes could be implemented in any of the villages served by JEMS or other communities outside the JEMS area in collaboration with another development partner agency. It utilizes popular theatre and participatory approaches and practices to mobilize residents to participate in community development strategies and programmes (Chambers, 2002). The cultural wake is a festive event designed to promote unity, promote traditional art forms and practices and allow community residents to display their talents and creativity. It also fosters collaboration among stakeholders and draws on diverse sources of knowledge to enhance the adaptive capacity of communities (Leach, 1996; Butler, 2015; Simmons, 2018, 2020).

3.4.6.4 Sports Development

JEMS utilized sports as a tool for development (https://pdf.usaid.gov/pdf_docs/PNADE352.pdf). The sports programme aims to foster a competitive spirit, partnership, cooperation, friendship and a network system within and beyond the villages. Sports are used at two levels to promote the above values: (a) the organization would organize stand-alone sporting festivals among the villages in softball cricket, five-a-side football, netball, tug of war, domino matches and other games to create competitive spirit among residents as well as mobilizing them to collectively address a fundamental development challenge in the community and (b) mobilizing resources to sponsor teams of players from the villages to participate in competitive sports at district and/or national levels, namely, cricket, soccer and netball. The JEMS sports programmes were very effective in enhancing the capacity of individuals and communities to resolve the development challenges within the villages (Rittgens & La Gra, 1991; Leach, 1996).

3.4.6.5 Community Education/Infrastructural Development

JEMS utilizes its community education programmes to enhance residents' awareness of the problems affecting their community and mobilizes them to participate as volunteers (providing their labour and skills on a self-help basis) to improve the knowledge, skills and infrastructure base of the community (Rittgens & La Gra, 1991; Leach, 1996). Through the programmes, information is provided to the community on the importance, roles and execution of each development project the organization is implementing in the various villages. JEMS has implemented over 200 sustainable projects over the past 45 years.

These include the Lower Stubbs Health Improvement Project, village water projects, the Enhams Development Project, community feeder roads, the construction of the community and emergency centre in Enhams, the National Annual Coastal Clean-Up Campaign and, further afield, the Vieux Forte Health Improvement Project in collaboration with ROUTES, an NGO in St. Lucia. Over the period under review, JEMS collaborated with the parliamentary representatives and the appropriate government agencies to provide electricity in communities where they were not in existence. These community development infrastructure projects enhanced the health, sanitation, sustainability and economic well-being of thousands of residents of these communities.

3.4.6.6 Environmental Awareness and Climate Change

The mobilization of residents at the community level to ensure the protection of King's Hill Forest Reserve and the Milligan Cay Wildlife Reserve and Bird Sanctuary and the stopping of sand mining on the beaches of Argyle, Stubbs, Brighton and Diamond to protect the endangered turtles and other species were the first wave of JEMS environmental programmes. The second set of JEMS educational programmes focused on strengthening the capacity of communities to develop their own institutions to enable them to implement sustainable livelihoods and development programmes within their respective communities. The SEDO organization in the village of Stubbs evolved from this programme. Therefore, JEMS has been part of the solution for achieving sustainable development practices by promoting community-based actions and policy changes.

Through support from UNEP, JEMS created the Caribbean Youth Environment Network (CYEN) in 1992. CYEN is currently working in 16 Caribbean countries to implement the JEMS participatory sustainable development model of building the resilience of young people and their communities to adapt to the impact of climate change (https://cyen.org/). Through the leadership of JEMS' founder, Andrew Simmons, the

International Youth Development Network (IYDEN) was created in Geneva, Switzerland, at the Third Preparatory Meeting of the UN Summit on Sustainable Development (UNCED) in 1991. IYDEN effectively coordinated the youth forum at the Earth Summit in Brazil in 1992. Wagaki Mwangi, chair of IYDEN, presented a speech on behalf of world youth to the heads of governments in the UNCED (Third World Resurgence, 1992). Although IYDEN is currently dormant, it effectively coordinated numerous global environmental activities for and with youth across Africa, Asia, the Caribbean and the Pacific region with the support of the Commonwealth Youth Programme, Commonwealth Secretariat, UK.

A critical aspect of JEMS' work is its emphasis on developing programmes to reduce the impact of climate change at the community level by implementing popular educational awareness programmes (Simmons, 2021), tree planting and national clean-up programmes in villages and schools across St. Vincent and the Grenadines. JEMS is not only able to challenge national policies that could harm the environment but was able to provide direction and support for dealing with these at the community and national levels (https://www.goldmanprize.org/recipient/andrew-simmons/). The organization won the Caribbean Conservation Award in 1993, and its founder, Andrew Simmons, received the UN Global 500 Award in 1990 and the Goldman Environmental Prize/Award in 1994 for outstanding contributions to the protection and conservation of the global environment (https://www.goldmanprize.org/recipient/andrew-simmons/).

3.4.6.7 Other Development Programmes

Sustainable Agriculture

As early as 1988, JEMS embarked on developing a cooperative sustainable farm project in Diamond, southeast St. Vincent and the Grenadines, on enhancing residents' livelihoods. It employed the services of a trained agronomist, Paddy Thomas, to work with small farmers in the 15 villages to enhance their agricultural, management and soil conservation practices in these communities. The organization leased 10 acres of land from the Diamond Diary Organization, the Government of St. Vincent and the Grenadines, to establish a model farm project. The farm was also used as a demonstration site for training farmers in new agricultural practices and to prepare emerging farmers to enhance the management of their own small farms. Through support from the Organization for Rural Development (ORD), JEMS embarked on a turmeric production project (3–4 acres); however, when the product was ready to be harvested, the market collapsed and was no longer viable. The collapse of the turmeric cash crop pilot project saved the farmers the stress of investing in the crop the following year. Due to the piloting of the cash crop at the JEMS farms, farmers from the community used the opportunity to observe the process and enhance their skills to grow the new crop. Therefore, the collapse of the turmeric market did not affect them because the JEMS farm buffers the project's full impact on the farmers, while farmers across the country were struck by the devastating failure of the project (Rittgens & La Gra, 1991; Simmons, 2021,2022).

Disaster Climate Risk and Vulnerability

Due to the location of the villages in the southeast of mainland St. Vincent and the Grenadines, preparing villagers to build resilience to natural disasters has always been a major focus of JEMS work. This programme consists of educating villagers on how best to avoid loss of property in periods of hurricane risks and vulnerabilities, the setting up of village emergency centres and the training of carpenters, masons and tradesmen to build houses that can withstand winds of over 150–200 miles per hour. In 1988, Hurricane Emily supported over 100 houses in the villages. JEMS was responsible for mobilizing 150 builders and building materials to rebuild houses. JEMS collaborated with the Ministry of Public Works and the social investment funds (SIF) to renovate and convert the community centre at Enhams into an emergency relief centre to shelter persons from national disasters (Rittgens & La Gra, 1991; Leach, 1996).

HIV/AIDS, Drugs, Crime and Violence

JEMS embarked on a Peers Education Approach to build awareness and reduce the impact and incidence of HIV/AIDS and sexually transmitted diseases among young people and residents in the villages. The project empowered young people to avoid engaging in illicit sex and risky behaviours. The project was funded by USAID and supported by US Peace Corps volunteers. The youth utilized the skill acquired from the project to work with other young people across communities southeast of St. Vincent and the Grenadines.

Communities in Stubbs and Diamond faced high gun crime and violence. JEMS created the Southeast Development Inc. (SEDI) in the Stubbs area, which is a CBO that utilizes steel pan music and cultural art practices as the mechanism to engage young people in community development programmes as a strategy for reducing the impact of violence, drugs and crime in the community. The SEDI initiative was initially funded through a government social investment fund, the Pan Against Crime Initiatives and the Embassy of Canada Small Grants Programme. SEDI participated in similar development projects in communities encountering high incidences of crime and drugs in St. Vincent and the Grenadines. It also placed third in the National Carnival Panorama Competition.

3.4.6.8 Lessons Learned from the JEMS Case Study

Numerous lessons are learned from the bottom-up participatory approach implemented by JEMS to achieve sustainable development within communities across St. Vincent and the Grenadines. These lessons include:

- The JEMS development strategy is based on a bottom-up approach focused on putting people at the centre by ensuring that they are engaged in shaping and influencing their own development (Leach, 1996; Angelsen et al., 2011; Danielsen et al., 2009; Commodore et al., 2017; Eicken et al., 2021).
- JEMS' work has demonstrated over the past 45 years that the bottom-up participatory development model possesses the potential to build community cohesion, foster empowerment and enhance sustainability, whereby people take responsibility for determining their own destiny. The model also includes the possibility of strengthening greater adaptive capacity by ensuring sustainable development (Chambers, 2002; Hernandez Delgado, 2015; Butler, 2015).
- There is a need for JEMS to find creative ways to attract and secure a sound financial base so that it can become sustainable and viable in the long term. It is challenging to depend on voluntary contributions to implement such vigorous and sustainable development programmes. The organization must focus on building sustainable financial resources, for example, creating and investing in small business development and enterprises to ensure the availability of sustainable resources to finance its programmes (Fox & Schearer, 1997).
- Over the years, JEMS has devised appropriate strategies on how to protect the environmental resources in the community while at the same time working with the residents to enhance their well-being. Therefore, JEMS should continue exploring innovative strategies to utilize these environmental resources so that residents within these villages can create sustainable livelihoods without destroying these valuable resources.
- While migration has taken its toll on the development of the communities and the country, JEMS continues to maintain a cadre of over 1000 volunteers who can call at any time to participate in the organization's programmes. Maintaining a solid base is critical to the organization's longevity and ability to remain a constant and effective agent for change in the community. Therefore, as people continue to move out of the country to seek greener pastures, JEMS would need to find ways to maintain committed membership and leadership; hence, capacity-building is important (Enekmark, 2003; Khan et al., 2018). Efforts should also be made to commit individuals who have already benefitted from the organization to give back some of what they have

personally gained to support the organization's work.

- Adapting to the changing social, economic and political landscape in St. Vincent and the Grenadines is vital to the survival of organizations such as JEMS. Hence, there is a need to constantly adapt, reshape, strategize and keep abreast of the changing environment internally and externally.
- JEMS has demonstrated the effectiveness of using culture (cultural wake) as a development tool through music, dance, drama and songs to mobilize, strategize and create awareness of the community's developmental challenges and empower the residents for action. No other organization has demonstrated the impact that cultural art forms and expressions can have on achieving Sustainable Development Goals. To its credit, the "cultural wakes" popularized by JEMS have highlighted the effectiveness of a cultural approach to development while at the same time preserving the country's rich cultural heritage and tradition (Rittgens & La Gra, 1991; Leach, 1996; Simmons, 2021). A similar lesson is learned for JEMS "Sports as a tool for development".
- The knowledge base of residents of the villages on climate risk disaster management was enhanced. This shows that the community education and awareness programmes implemented to enhance the skills and knowledge of residents of the villages to avoid loss of lives and property in the event of natural disasters were very effective.
- The mobilization of skilled volunteer workers by JEMS to undertake the rehabilitation of houses and other public facilities destroyed by storms in the villages was very effective in (a) reducing the cost to implement such projects and (b) decreasing the incidence of corruption and mistrust among government workers and the residents of the community.
- Women and youths were deemed the most vulnerable in the community and were readily mobilized to contribute to the community's overall development. The enormity of the involvement of women in the projects was acknowledged as the primary factor responsible for the successful implementation of the projects. The men were motivated to participate in the projects due to the many women working tirelessly on these projects.

3.4.6.9 Best Practices Evolved from the JEMS Work

After more than 40 years since its founding, JEMS has adapted and responded to the changing development environment in which it is based. While the organization's issues have remained constant, the economic, political and social landscape in St. Vincent and the wider Caribbean has changed dramatically. Consequently, JEMS has had to constantly reposition itself to meet these development changes.

Over the years, JEMS has embraced changes with the same enthusiasm it came into existence in 1978. Having evolved from an agenda focused on providing basic needs, the emphasis has shifted to enhancing people's quality of life and capacity to respond to changes in sustainable lifestyles and innovative pathways. JEMS continues to ensure that it has a programming agenda that is flexible and relevant to the community's development needs and is assisting these communities in building resilience to enable them to respond adequately to emerging development challenges such as climate risks and disaster vulnerability.

JEMS has set itself apart from similar development entities by developing a bottom-up participatory integrated sustainable development approach centred on the people, the actual designers, implementers and beneficiaries of its programmes (Rittgens & La Gra 1991; Leach, 1996; Chambers, 2002). The road ahead will be filled with challenges, but one sure thing is that JEMS has built a development model capable of managing these challenges.

- The project contributed to developing strong and dynamic local leadership within the villages. The local leadership has been strengthened through the development and operationalization of village emergency committees (VEC), which are responsible for managing the operation of disaster emergency shelters (DES) and preparing residents to take precautionary measures to protect themselves and their families in times of natural disasters.

- JEMS collaborated with the government's National Emergency Management Organization (NEMO) to establish and train the committee members to adapt to the impact of climate risk and disaster challenges facing the community. Such collaboration facilitates the strengthening of partnerships among all key stakeholders.
- The development and implementation of innovative community-integrated participatory approaches contributed to building the capacity of the community to enhance project management, monitoring and evaluation within participating communities. JEMS utilized the "cultural wakes" and aided "self-help" approaches to raise residents' awareness of the severity of the development challenges facing these communities and mobilized volunteers to implement actions to enhance the community's infrastructure, skills and knowledge of residents.
- The fostering of collaborative and cooperative spirit among grassroots people in villages as a strategy for solving development challenges. JEMS was very effective in providing leadership to mobilize a range of government agencies, civil society and business partners to work together to provide resources to aid in the construction of villagers and feeder roads, enhanced housing stocks, education and sanitary facilities and pipe-borne water in these communities.
- Youth and women possess the capacity to make positive contributions to the realization of positive development outcomes at community levels, specifically in the areas of climate risk and disaster management. The young women and single-parent mothers were responsible for leading the initiative to establish district emergency centres and committees and engage in capacity-building training, networking and actions to resolve the community's vulnerability challenges.

3.4.6.10 Recommendations

JEMS has demonstrated how a CBO can be involved in developing strategies and programmes to empower local communities to conserve critical environmental resources and influence policies and programmes at the national level to achieve sustainable development. It successfully conserved King's Hill Forest Reserve and the Milligan Cay Bird Sanctuary and put the spotlight on the impact of climate change on small developing island states (SIDS). However, it needs to continue to collaborate with partner agencies to develop initiatives to utilize the eco-tourism sites at Rawacou Beach; turtle-nesting beaches at Brighton, Stubbs and Diamond; and the Argyle International Airport to ensure that residents at the village level benefit from these resources to enhance the sustainable livelihoods of the community. Therefore, its unstable financial situation would determine that the organization moves in this direction.

JEMS' record of collaboration with government agencies and the NGO community, other partners and stakeholders stands as a testament to the respect the organization's work is held and the impact it has had on national development. However, the organization needs to put more effort into getting government and development partners to "buy in" and support the long-term notion of conserving and preserving the natural resource base of these communities in sustainable ways.

Investing in capacity-building to enhance conflict resolution and mediation skills is a worthwhile initiative that can help temper some of the apparent partisan perceptions often associated when non-governmental organizations and governments are at variance on issues of sustainable development. In fact, it is a well-established fact that JEMS' collaborative linkages with its numerous partners and stakeholders have been pivotal to the organization's growth, development and sustenance. One of its key strengths is its robust collaborative links and cooperation among stakeholders and partners.

On the issues of climate risk and disaster management, JEMS on the mobilization of the local community to rebuild the housing stock and basic infrastructures in the villages in the southeast of St. Vincent and the Grenadines shows that the organization possesses an immense capacity to build self-reliance, through the mobilization of community people to empower themselves to address the fundamental development challenges affecting their communities. For the organization to continue to build resilience at the community level to effectively manage climate risk and disaster vulnerabilities, JEMS needs to continue to focus its resources on the development of the capacity of residents and key stakeholders to enhance and sustain the skills and competencies gained and the lessons learned during the implementation of the projects by:

- Ensuring that the district emergency committees (DECs), which were created to manage the emergency centres and to prepare the residents to take precautions to save lives and withstand such challenges in the future, continue to function effectively.
- Targeting and mobilizing resources (financial and technical) to ensure the organization could provide such development support to people and their communities in similar circumstances in the future.
- Mobilization of resources to ensure the continued implementation of climate change and sustainable development training programmes for community leaders and key stakeholders to work with communities to build their resilience to adapt to the impact of climate change and other development calamities.
- Provide training for carpenters and masons involved in the construction industry to build houses to withstand winds of over 150 miles per hour.
- Continue to lobby the Physical Planning Division in the Ministry of Planning and Finance to affect the building code to ensure houses are built to national standards and enforce these codes and policies.
- Due to the limited financial resources available to implement its programmes, JEMS should ensure that it prioritizes the mobilization of sustained financial resources to support its climate change and sustainable development programmes.

3.5 Case Study 2: Top-Down Approach Implemented by the Community Development Division in the Ministry of Social Development, Government of St. Vincent and the Grenadines

The second case study explores the strategies implemented by the Government of St. Vincent and the Grenadines Community Development Division (CDD) in the Ministry of Community Services from 1990 to 2000. The poor performance of the top-down development approach implemented by the CDD resulted in increased poverty in rural and urban communities. This prompted the government to install a new minister and recruit a new director of community services to manage the operations of the Ministry. It also resulted in shifting from a top-down to a hybrid participatory integrated development approach which increased the participation of communities in the development and governance of the programmes.

The hybrid approach consists of a combination of top-down and bottom-up approaches and participatory practices when implementing basic needs development programmes at the community level, drawing from the strengths of both approaches (Matland, 1995; OECD, 2013; Stockowiak et al., 2016). Although the hybrid approach was questioned by policymakers and senior technicians in the Ministry in its earlier stage of development, it was later hailed by these policymakers and community leaders as an incredible success based on (a) the numerous amounts of community volunteers who mobilized and participated in the "aided self-help" projects, (b) the increase in the number of young men and women graduating from skills development,

(c) the number and quality of training and community education programmes implemented, (d) the high quality of the projects prepared and submitted by communities and (e) the increase in the number of projects successfully implemented by communities.

The hybrid approach resulted in the increased mobilization and participation of local communities across St. Vincent and the Grenadines in programmes to address skills shortages, enhance knowledge, build on the development challenges these communities face, improve community infrastructure and enhance sustainable livelihoods. The case study captures the achievements, gaps and pitfalls of the Community Development Division between 1900 and 2000 to implement capacity-building initiatives to enhance the competence and capability of the division to deliver sustainable development programmes across communities in St. Vincent and the Grenadines. The case study highlights the achievements, gaps and pitfalls and makes recommendations for enhancing the CDD capacity-building programmes in the future.

3.5.1 The Socio-economic Perspective of St. Vincent and the Grenadines (SVG)

SVG is described as a small low-middle-income economy exposed to external economic shocks and natural disasters (CARIBSAVE, 2012; IMF, 2017). The country is plagued with a high incidence of unemployment and poverty resulting from the lack of appropriate skills in the workforce for the jobs that are available in the market. Only 5.4% of the population has attained tertiary education (CARIBSAVE, 2012). Economically, agriculture and tourism are the main economic earners of the country, accounting for 35% of the GDP, respectively. Both sectors are experiencing severe reductions in economic growth due to the destruction of the banana cash crop by invasive diseases and the global economic crisis. The resulting impact of such economic difficulties has resulted in a decrease in economic growth from 5% in 2000–2001 to below 1% in 2017 (IMF, 2017). The economic situation in St. Vincent and the Grenadines was worsened in 2021 by the COVID-19 pandemic and the eruption of the La Soufrière volcano.

The economy of SVG is not diversified, as approximately 75% of its economic output is derived from two sectors, agriculture and tourism. The global economic crisis resulted in a fall in per capita GDP in 2010–2011. The economic problem was further compounded by the destruction of the banana industry by invasive species such as Moko and Black Sigatoka diseases, which have contributed significantly to the reduction in the number of persons earning their livelihoods from the agriculture sector. Women and youth were severely impacted by this devastating event. Before its demise, the banana industry brought over $150 million per annum into the economy of SVG and was responsible for employing approximately 20,000 persons (informant – former director of agriculture). The economy of SVG is vulnerable to external and internal shocks and stressors. Further to losing the European market for bananas, the country was struck by a series of natural disasters; sluggish global demand and slow implementation of key infrastructure projects further hampered the economic recovery of St. Vincent and the Grenadines from the global financial crisis (IMF, 2016).

Poor income distribution and high poverty continue to plague SVG's women and youth. A poverty study conducted under the Country Poverty Assessment 2008 (CPA) revealed that between 2007 and 2008, the poverty headcount index was at 30.2%, a reduction of 7.3% over the 1995–1996 figures (Government of St. Vincent and the Grenadines, 2008). The "rural banana farming population was identified as being at high risk and needed particular poverty reduction strategies" (Government of St. Vincent and the Grenadines, 2013a, b, pp. 44–5). The poverty report brought out the growing inequality in well-being and the lack of resources injected into rural and urban communities after the financial crisis in 2008–2010 due to the demise of the banana industry. The considerable increase in social welfare payments to the destitute and the poor indicates the worsening poverty situation in the country (Government of St. Vincent and the Grenadines, 2017; Simmons, 2018). The IMF (2017) country report on SVG indicates that unemployment rates are estimated at 25% of the workforce and have reached 46% among the youth (15–24 years of age). The report states that the unemployment rate is high, particularly for the youth, the less educated and women. While there has been an increase in the participation of women in the labour market from 45% to 65%, unemployment among women rose from 18% to 30% between 2001 and 2017.

The Government of St. Vincent and the Grenadines has utilized a substantial amount of its resources to the repairing infrastructure, particularly after calamities such as storms and hurricanes. It has also used its limited resources to build prestigious projects, increasing the national debt burden (IMF, 2017). These projects also require high maintenance support, which is an additional increase in the national debt. St. Vincent and the Grenadines is exposed to climate change and other development calamities impacting negatively on SIDS (IPCC, 2014a, b, c, 2022; Simmons 2018). Climate disaster risk management issues are one of the main development challenges plaguing the residents of rural and urban communities in St. Vincent and the Grenadines. Hurricanes, storms and flooding are these communities' most dominant challenges.

3.5.1.1 The Population of St. Vincent and the Grenadines

The country's population is predominantly of African descent and is estimated at 104,332 (51% men), of which 33% are estimated as young people (World Bank 2021, https://data.worldbank.org/indicator/SP.POP.DPND.YG?locations=VC), comprising 56,419 males and 53,572 females, of which 91.9% live on mainland St. Vincent. Over 45% of the population is estimated to be young men and women. The island's mountainous interior has forced 90% of the population to live on a narrow coastal belt of land 5 meters high and 5 miles inland. Due to the high population density, housing settlement areas are now extended on lands designated for agricultural purposes, on slopes of central mountain ranges, along hilltops and in old riverbeds (CARIBSAVE, 2012). The practice has created several difficulties that are detrimental to the environment, including:

(a) Increased deforestation and runoff water which has resulted in landslides and flooding for persons living downstream.
(b) Reducing the penetration of rainwater into the soil.
(c) Decreased food security by reducing the amount of land available for farming.
(d) Increased pollution of streams and coastal areas.
(e) Exposure of the poor, marginal and most vulnerable members of society, mainly the youth and women, to live in areas that are vulnerable to climate events because they do not have the resources to purchase land nor build sturdy houses to withstand the intensity of the storms and other climate stressors (Simmons, 2018, 2021).

This problem is compounded by the limited adherence to physical planning and land management policies and laws within communities.

3.5.2 Overview of the Community Development Division, Ministry of Housing and Community Services

The CDD was one of the eight divisions in the Ministry of Social Development, now referred to as the Ministry of National Mobilization. According to the website of the Ministry, its purpose is to engage in social transformation through social empowerment, social protection and justice, using national mobilization, social development, youth and sports. During the period under review, the Ministry also has responsibility for (a) cooperative development, (b) housing, (c) local government and the Kingstown Board, (d) gender affairs, (e) social welfare, (f) sports affairs and (g) youth development.

The mission of the CDD was to facilitate social, cultural and economic development at the community level. The agency's mandate is to "assist Community Based Organisations, Non-Governmental Organisations, and other stakeholders to build an enabling environment for communities and marginalised people to participate actively in shaping and influencing national development" (http://mobilisation.gov.vc/mobilisation/index.php/community-development-division).

The Government of St. Vincent and the Grenadines views community development as a model for achieving sustainable development intended to significantly affect self-sufficiency and high quality of life.

3.5.2.1 CDD Links and Partnerships with Other Organizations

The CDD is a government agency based in the Ministry of Social Development. Initially, before 1990, the division implemented a top-down approach to develop and implement its programmes at the community level. Informants posit that the division's work was poorly coordinated, compounded by poor leadership, unskilled staff with low capacity and competencies and limited comprehension and skills regarding applying novel community development strategies. This resulted in a shift of the CDD approach to a hybrid development approach which empowers people to take action to build their resilience to address the climate risk vulnerability and "basic needs" challenges facing these communities (Matland, 1995; OECD, 2013; Stockowiak et al., 2016).

Before the period under review, the funding to CDD was mainly used by the Ministry to implement infrastructure programmes in a limited number of constituencies, mainly community centres. Most of the community centres were in dilapidated conditions and were not fit for use by the communities. Besides lacking basic infrastructure, most communities lacked pipe-borne water, un-paved roads and sanitary facilities and preschool and daycare facilities.

The country was encountering an acute squatter settlements problem on the agricultural estates the government acquired as part of the government's Land Reform Programme. The implementation of the land distribution programmes was slow, was disorganized and lacked transparency, so people became impatient and began squatting on these agricultural estates. The squatting events created "ghetto"-like communities, which began to fester into increased poverty, crime, gangs and deforestation.

The new strategic programme instituted by the CDD, accompanied by a hybrid participatory integrated development approach, was approved by the government's Cabinet (Matland, 1995; OECD, 2013; Stockowiak et al., 2016). The Government of St. Vincent and the Grenadines repositioned the division as a central mechanism of the government's "war" on poverty and underdevelopment by incorporating the division's role in national development planning. The CDD was defined as a catalysis for the effective development, implementation, distribution and delivery of government resources to communities. The division collaborated with a range of development partners, including civil society and private sector agencies, to ensure equitable distribution of resources and enhance the participation of communities in the governance of their communities.

3.5.3 Strategies Programme Implemented to Build the Capacity of Communities

The CDD utilized numerous strategies to build the capacity of people and their communities. The capacity-building strategies implemented by CDD were borrowed from JEMS's participatory integrated development approach (Simmons, 2021). The new director of community services was affiliated with the JEMS organization and restructured the top-down approach implemented by CDD into a hybrid approach to ensure smooth implementation of the approach into the governmental institution (Matland, 1995; OECD, 2013; Stockowiak et al., 2016). This capacity development strategy implemented consists of training, provision of technical assistance, sharing of innovative knowledge, networking and coaching. Although the policymakers were initially sceptical about adapting to elements of the bottom-up participatory integrated development approach introduced by the new director, they were motivated to approve the new approach due to the communities' positive response to the CDD programmes' innovation. This was also compounded by the following: (a) the current community development strategic approaches and programmes were failing these communities, (b) the people did not have faith in the programme implemented by the division, (c) the government was under severe pressure from the DfID (external funders) to change the approach of the division's programme to respond to the felt needs of the communities and (d) the policymakers and communities were fully aware of the benefits to be derived from the new approach as they witness the transformation experienced in the communities served by JEMS.

The innovations in the CDD programmes were as follows:

Cultural Wake Programmes The cultural wake programmes utilized music, arts and culture and other participatory approaches to raise awareness of communities on the problems affecting the communities and motivate the people to participate in programmes/projects within the community to resolve the development challenges confronting the community (Chambers, 2002). The processes involved in developing and implementing cultural wake activities were defined in the JEMS case study (Leach, 1996; Rittegen & La Gra, 1991). Community development officers (CDOs) implemented the cultural wake initiative throughout the communities across St. Vincent and the Grenadines. Stipends and transportation support were provided to Naked Rooks Drumming Groups and other cultural groups to provide musical accompaniments to communities to support the performers during cultural wake activities. After the artists performed their calypsos and poetry in the village square on specific themes, a community animator will engage the community in discussions on the issues affecting the community, the extent of the challenge and solutions to resolve the development challenges facing the community.

3.5.3.1 Aided "Self-Help" Approach

> In Owia village, when it is raining, the road to farms will be flooded, and the farmers would not be able to go to their farms in the mountains for weeks. They would have to go all the way to Fancy, miles away, to get back home. CDD assisted them in building a bridge so they could access their farms when the river was flooding. The bridge provides access to and from the farms in approximately 20 minutes. The farmer worked together to build the bridge. One farmer brought their electric generators so they could use the electric drill to bolt up the rails on the bridge. Glaston Buddy Lavia, former CDO for North Windward.

This is one of the most innovative capacity-building development strategies implemented by the Community Development Division. The "aided self-help" approach incorporates the cultural wake with developing and implementing the basic needs projects. The community will be engaged in a series of activities leading to submitting projects to the Project Appraisal Committee (PAC) of the CDD. The PAC was approved by Cabinet and comprised representatives from civil society and governments. The role of the PAC concerns the assessment and approval of projects to be funded under the $2.3 BDD-funded aid funds annually. The Community Development Division provides funding in material support to communities between EC $1000 and EC $50,000 (US$1 = EC $2.7) per annum to support the implementation of projects in communities ranging from pipe-borne water; community and feeder road development; daycare facility and preschool, sanitary and recreation facilities; sports complex and hard courts; community centre; sustainable livelihoods; and education programmes. The division provided material and technical assistance, while the community provided labour and food for the projects. The division would coordinate the participation of skilled volunteers from other communities to participate in the projects. These volunteers include carpenters, masons, plumbers, electricians, etc., from different communities to work with communities across St. Vincent and the Grenadines. These skilled volunteers will be transported to and from projects on weekends.

Another aspect of the aided self-help programme is the provision of technical assistance to communities to prepare projects for submission to the Ministry and other donor agencies such as USAID, CIDA and GEF Small Grants Programmes. The US and Canadian Embassies in Barbados and UNDP directly provide funding through the CDD to support the implementation of projects within communities. After funding is provided to communities to implement development projects. The division endorsed projects provided to these external agencies and provided technical guidance and support to communities to ensure the implementation of the project.

A successful component of the division's work is the opportunity to work with a range of governmental ministries and divisions to assist them in providing the needed technical assistance to communities that need these resources. The Central Water and Sewage Authority (CWSA) negotiated with CDD to provide all pipes to communities at a 50% discount and technical support to ensure the pipe-borne water projects were properly implemented. After implementing these projects, the communities would have a cadre of trained plumbers available to earn a livelihood from implementing pipe-borne water projects for households in their respective communities. In the case of the Ministry of Agriculture, the agriculture officers from the Ministry will teach farming skills to farmers involved in the division's food security programme to enhance food and animal husbandry production within these communities. It will also work with the Red Cross to train residents to enhance their disaster climate risk and preparedness skills in communities.

Climate Risk and Disaster Management in Communities Although the government currently has a designated agency in charge of disaster risk management, the National Emergency Management Organization (NEMO) agency was not in existence during the period under review. The responsibility for disaster management was in the Ministry of Social Development under the Disaster Preparedness Unit. It was the responsibility of the community development officer (CDO) to work with the disaster preparedness officer to train and mobilize communities to build resilience to respond to hurricanes, storms, volcanic eruptions and other disasters. The CDD was responsible for retrofitting communities' public buildings to withstand hurricanes at over 150 miles an hour and working with community leaders to set up emergency committees and shelters within communities. Such capacity-building training consists of educating communities on the impact of natural disasters, developing community disaster plans and programmes and managing and operationalizing disaster emergency shelters. There were tensions regarding the roles of the Disaster Preparedness Unit, the parliamentary representative responsible for the areas affected and the emergency shelter management responsible for allocating resources to communities affected by disasters. The CDO would be the one who is burdened with the task of clarifying such roles with the community to ease such tensions.

3.5.3.2 Community Education

> The community education programme is implemented through the implementation of a series of workshops and or consultative meetings with the community on how the people could utilise the natural resources of the communities to develop themselves and the communities. In a situation where the high literacy rate is identified as a problem, the CDD would collaborate with the Adult and Continuing Education Division in the Ministry of Education to implement adult literacy programmes in schools, community centres, and/or churches. Roger Young, former head of the Community Development Division, Ministry of Community Services.

The community education programme is divided into three components:

(a) Training for community leaders and their organizations in leadership and governance. This component aims to enhance the leadership capacity of communities to resolve the development challenges facing these communities.
(b) Training of staff and other civil servants who are desirous of developing programmes at community levels. The staff of CDD is trained in social research, entrepreneurship and livelihood development, conflict resolu-

tion and community mobilization strategies. Training is implemented internally and externally every quarter. Some staff pursued studies at the University of Wales and the University of the West Indies, leading to a certificate in social welfare, a diploma in development and a masters' in economic development.

(c) Skills for employment and entrepreneurship development training which provide specific training in craft development, carpentry and masonry, auto mechanics and agriculture. The CDD operated over ten such centres in collaboration with community-based organizations (CBOs) and other governmental agencies, including craft development. These training centres train trained young men and women in skills to enable them to work with others to set up their own businesses. The training was provided by skilled, experienced professionals. In a situation where the specific skills were not available, experts were recruited through the US Peace Corps and VSO to manage the skills training centres for a period of 2 years, within which period the external trainers will be responsible for training local trainers to take over the centre (ILO, 2010; UNDP, 2015).

3.5.3.3 Skills for Employment

The skills for employment programme is a sub-programme of the community education programme. Many rural and urban youth and women in St. Vincent and the Grenadines drop out of school and are deemed illiterate and unskilled. To address this problem, the CDD collaborates with NGOs to develop skills training centres in designated communities to provide primary/functional literacy and numeracy skills as well as employment skills in the trades (plumbing, carpentry, masonry and auto mechanics). In collaboration with the Craft Department and the Taiwanese Technical Mission, the CDD operated craft development training centres in Kingstown, North Union, Barrouallie and Orange Hill. These skills training and craft development centres created employment for thousands of youths and women during the period under review through building construction, agriculture and tourism in St. Vincent and the Grenadines and across the Caribbean region.

3.5.3.4 Integrated Village Development (IVD) Programme

This programme is comprised of combining all the components of the CDD programmes in one designated area. The overall aim of the strategic programmes was to target the communities with a high incidence of poverty and empower the people with skills and knowledge to break the silos of poverty and underdevelopment. The IVD programmes were implemented in the communities of Fitz Huges (Chateaubelair), Glebe Hill (Barrouallie), Spring Village (Georgetown), Francois (Vermont), Bottom-town (Kingstown, the capital), Peruvian Vale and Gen. These areas were selected based on the high incidence of poverty revealed in the National Poverty Assessment Report (Government of St. Vincent and the Grenadines, 2008). Several skills for employment training, pipe-borne water, agriculture to enhance food security and sanitary facilities (toilets and public baths) were implemented in these micro-communities:

> People in these communities were happy because they no longer had to go into the road to bath nor to collect water at the public standpipes in these communities...the children did not defecate nor chucked the household garbage into the streams... they were able to have pipe borne water delivered directly to their houses. A female resident of the community of Spring Village, Georgetown.

A representative of the communities where the Integrated Village Development (IVD) Programme was implemented attested to the reduction of water-borne diseases and vermin, such as rats and mosquitoes, due to the improvement in health and sanitation standards:

> People at the community levels were viewed as moving spirits, i.e., they are slow to speak but are well motivated and ready to work as volunteers on community projects. Participatory mapping exercises were utilised to identify the major resources in the community and hazards such as drugs, violent crimes, and illicit dumping of garbage in Spring Village, Georgetown. We identify what the community can do on its own, what could be done

> in collaboration with the government, and what it could do with help from outside agencies. I remember vividly that people were not able to cross the river to go to the clinic, hospital, and school in Georgetown, which was approximately 10 min away. They must travel all the way down to Mount Bentick, which will add an hour to their journey. We build a bridge over the river to shorten the distance to the hospital, health centre, and schools. Glaston "Buddy" Lavia, former CDO for North Windward.

3.5.3.5 Cross-Cutting Programmes

The environment, gender and youth were viewed as cross-cutting programme issues when implementing the CDD programmes; CDOs' concerted efforts were made to mainstream environmental and gender issues in all programmes. Monitoring and evaluation were implemented every quarter to assess the CDD programme's performance and provide feedback on how well the programmes were performing and whether it was reducing the incidence of poverty in these communities.

3.5.4 Factors for the Success of the CDD Capacity-Building Initiative

The CDD efficiently worked with other governmental ministries and divisions by persuading them to provide resources and technical assistance to communities needing them. The work with the Central Water and Sewage Authority (CWSA) was significant. An MOU was signed with the CWAS indicating that pipes for all water projects would be provided to communities at a 50% discount rate. Also, CWSA agreed to provide its staff to work with communities on all water-borne projects. It had committed itself to working with the CDD to train a cadre of plumbers who would be available to assist the communities in implementing pipe projects and earn a livelihood from the new skills acquired.

The agricultural officers are committed to teaching farming skills to farmers participating in the CDD agricultural programmes. The initiative contributed to enhancing the nutritional needs of communities as well as enhancing the livelihoods of residents.

The clarity and decisiveness of the leadership of CDD and the bold commitment of staff were attributed to the success of the Community Development Division's project. The staff was recruited from the communities; therefore, they knew the communities regarding the development challenges they encountered. The people in these communities also knew, respected and trusted the staff of the CDD:

> They knew where they lived and how to get in touch with the officers. It is important to note that most of the officers live in the Communities in which they serve. They know our role and the function in the community and how to get in contact with us when they need our help to assist them in resolving their problems. Learie Johnson, former community development officer.

The CDD organized staff exchanges, internships, coaching and apprenticeship programmes to build the capacity of staff and beneficiaries of the CDD programme. New staff was assigned to work alongside more experienced staff members in the field. This strategy effectively built the capacity of staff to deliver the projects at a community level. It also contributed to building the confidence of weaker staff and provided opportunities for stronger staff to build their capacity through mentorship and coaching. The strategies effectively facilitated the transfer of skills and knowledge to the community's residents. Reverse mentorship was also practised by encouraging new staff with skills in ICT to train the older and seasoned staff members.

Due to the intensity of the CDD programmes, the director designated a "mental day" once per month. This day was set aside for all staff to work with an experienced CD officer in their respective district on an innovative project. After field exercises are completed, they will move to a beach where they will participate in recreational activities with the community. The "mental day" initiative has unique advantages: (a) staff are provided with the opportunity to socialize with each other by working with other staff in different settings outside of the office and the community they are assigned, and (b) they have the freedom to respond to difference and/or variance in development challenges by applying a range of participatory development methodologic practices in a safe setting in the field, and (c) it also provides opportunities for staff to resolve their difference, building friendship and partnership among each other staff and members of the community in which the exercise is taking place.

In summary, there is a partnership between the staff of the CDD and residents of the community level. This situation is compounded by many people in communities who are prepared to work as volunteers in the CDD programmes. This was evident by the high motivation of staff to accomplish the tasks at hand were attributed to the "mental day" initiative.

3.5.5 Challenges and Risks Faced by the Organization and Communities

During the first year of implementing the new strategic direction of the CDD, the programme was re-energized, and there was a new wave of positive energies flowing within the communities. The communities began setting up a wide range of community organizations to access the technical and financial resources available to the communities. The demand for support from the CDD far outweighed the technical and financial resources available. The CDD implemented over 100 projects annually within the community across St. Vincent and the Grenadines. Therefore, by year 3 of the implementation of the CDD project, the financial resources available to fund the programme were deemed very limited to support the high demands of the CDD programme:

> With the high level of momentum and demand for community projects, the CDD could have been able to do more work if it was given more money. Amor Jackson, CDO for North Windward South Leeward.

The high demand for the CDD programmes and the complexities of the communities' development challenges required highly trained professional staff to resolve these communities' development challenges. Some members of the CDD did not complete their secondary education. Therefore, they lack the minimum requirement to be employed full-time in the civil service of St. Vincent and the Grenadines. The limited number of staff of the CDD who accessed tertiary-level education was promoted outside of the CDD to other ministries in the civil service due to the limited number of posts available in the CDD to accommodate their qualification. This resulted in reducing the number of trained professional staff in the CDD.

The morale of staff was severely affected by the temporary nature of the employment of the staff. Compared to the agricultural extension and the public health officer status, the CDO felt they were not properly paid for the work they were delivering to communities:

> The staff were trained in a limited way. I believed that the Heads of the Ministry did not support the training of staff. Alpha Williams, former community development officer.

The CDD has an unreliable fleet of old jeeps and trucks that would break down frequently due to the poor service provided by the garage of the government's Ministry of Public Works. The poor operation of the transport service created delays and bottleneck in the field. This situation was compounded by business houses in Kingstown; the capital refused to deliver products to communities to implement the project as the Government Treasury Department did not pay them promptly. Although CDO was required to work with communities to complete their project proposal, the quality of documents submitted to the CDD was poor, thus resulting in frustrations within communities. Initially, the communities could not develop proper fundable project proposals to access funding from the CDD and other donor agencies.

Political interference from ministers and senior government officials was a serious issue affecting the smooth implementation of projects in communities. There were some instances when the minister of community services would delay signing the MOU to access the funding from DfID due to his dissatisfaction with the number of projects approved in his constituency. Such delays in implementing the CD programme could be up to 6 months.

3.5.6 Strategies Implemented to Address the Capacity Development Challenges

Regarding the issues pertaining to low status, poor payment and the temperature nature of staff employment, the director of community services took the staff to the Office of the Director General of Finance and Planning to protest the situation. This situation was compounded by the refusal by the minister to sign the MOU with the DfID, which resulted in the delay of staff salaries for 6 months annually. This put additional pressure on staff to meet to adequately respond to their family's well-being. The outcomes of the protest action include (a) appreciation by the director general of finance of the contribution of staff of CDD to the achievement of the government's national development goals, (b) provision of a contingency fund to pay the salaries to staff and operationalize CDD programmes in situations where there are delays in the release of the DfID aid funds and (c) the promotion of two members of staff annually to the permanent staff of the civil service. The decision to promote two staff annually to the public service was welcome by the staff of CDD as a positive victory. They posit that the decision will impact the staff pension and other benefits.

Regarding the issue pertaining to the delay of the DfID aid fund, the CDD staff would meet with community leaders to explain the causes of the delays in funding. In situations where the communities were frustrated and threatened not to participate in the implementation of the project, the meetings reduced tensions between the communities and the government. They contributed to strengthening the relationship and partnership with the community. Tensions were constantly brewing with the parliamentary representatives regarding implementing projects in the communities. The CDOs frequently meet with parliamentarians to update them on the CDD project selection and approval process, the nature of the organizations from the community submitting the projects and the projects approved. This contributed to easing the tensions between the parliamentarians and the members of their constituencies and enhanced better relationships among all stakeholders.

3.5.6.1 The Methodologies Implemented to Enhance Capacity Development

The innovative participatory methodologies utilized by CDD to build the capacity of communities and the staff were very effective in mobilizing people and their communities to participate in the development and implementation of the projects, as well as enhancing the delivery of the government's poverty alleviation programmes to the community. The central government elevated and recognized the CDD's role in national development. The director of community services participated in numerous high-level governmental missions at regional and international levels, including the CDB UN Commission of Latin America.

The participatory methodologies gave the people a sense of ownership and belonging. The people were empowered because they were central in identifying their problems and developing solutions to the communities' development challenges (Chambers, 2002). The shifting of the role of CDD from a top-down development approach to one that is process-oriented combines both top-down and bottom-up approaches where the people are empowered to lead the type of development they want to take place in their communities. The CDO performed the role of facilitator and guided the development process, ensuring that the project implementation was on the right track:

> Yes, the methods were effective. It strengthens how I deal with people and how I handle situations in communities. It also helps me to work with people to work out their differences, plan the next steps and approaches to addressing the problems they are facing. Denise Harold, community development officer for Northern Grenadines.

The informants said that the participatory community mapping and planning methodology was one of the most innovative and effective strategies implemented in the communities. This strategy efficiently empowered the people to identify the problems in communities and develop projects and programmes to resolve these problems. The CDD seed fund provides an opportunity for direct interventions in the community to resolve poverty challenges. The innovation strategies implemented by the CDD opened access to the injection of external funding from donors to support the implementation of community projects. These donors provide funding directly to communities to support the implementation of environmental, health and sustainable livelihood projects.

There were instances where CDOs were experiencing difficulties preparing estimates for small-scale infrastructure development projects in communities. The CDD collaborated with the government's technical college and the Ministry of Public Works Technical Division to implement a series of workshops to enhance staff competencies in estimating buildings and other infrastructure projects. The training outcomes enhanced the estimation of the materials and improved the project proposals submitted to the Project Appraisal Committee. The staff also utilized the skills gained to conduct damage assessments at community levels during climate risk disaster events.

In summary, capacity-building strategies implemented by CDD to resolve the poverty dilemma in communities through the skills for employment training, aided self-help initiative and entrepreneurship development initiatives were successful. Youth and women were trained in entrepreneurship in the business skills training centres in urban and rural areas. The training enables graduates to develop skills in business plan development and strategies for set-up businesses. Seed funds were provided to unemployed youth and women to establish their own businesses. CDD was very effective in supporting the transfer of bookkeeping and accounting skills to business owners in communities, enhancing their businesses' efficiency.

3.5.7 Lesson Learned

The lessons learned are based on the notion of repeating the positive aspects of the project without replicating the mistakes of the project. This section captures the lesson learned, including (a) why it was relevant for the project and (b) how it should be applied to future projects. These lessons were as follows:

- Information was gleaned on the origin of communities, how the communities in SVG were developed and the provision of ideas for developing such communities in the future. This information was relevant in understanding the nature and history of these communities, implementing the demarketing process and devising sustainable strategies for resolving the current and future development challenges facing these communities.
- The implementation of participatory methodologies such as cultural wake and community mapping strategies contributed to the involvement and participation of the people in the development planning processes implemented in the community. These strategies give the people a sense of ownership, commitment and stake in developing their community.
- Educating the people of the communities on the challenges facing the community would enhance their participation in the overall development of their communities. This process empowers people to take on leadership and critical development roles in the governance of their communities and ensure sustainable development measures are adhered to.
- People would volunteer their skills and labour in programmes/projects to develop their communities when they are involved in deciding what development they would like to see taking place in their communities and know how they will benefit from these projects/programmes.
- People know the problems they are encountering, and they also have the answers to resolve these problems. Therefore, governmental agencies should implement a hybrid approach

(combining strengths of top-down and bottom-up approaches) to empower them to take the lead in solving the development problems in their communities.
- The communities lack the finances and technical expertise to resolve their problems. Therefore, besides providing finances through the Aided Self Grant Funds to communities to implement basic needs projects, the CDD's role is to build a partnership with other governmental and donor agencies such as the CDB Basic Needs Trust Funds, GEF SGP and SIF to channel development finance and technical assistance to communities which need an injection of resources to support their development.

3.5.8 Best Practices

Best practices are defined as guidelines, ethics or ideas representing each project's most efficient or prudent course of action (https://warontherocks.com/2019/02/in-defense-of-best-practices/). The CDD develops innovative practices that stand out as guidelines for developing and implementing community development programmes across the Caribbean. Several best practices emanating from the CDD are as follows:

- The utilization of participatory practices as a strategy to enhance participation and empower the poor and poverty-stricken living in squatter settlements and marginalized rural and urban communities to transform these squatting settlements into fully independent communities with basic infrastructure, including water, paved roads, and sanitary facilities, and communities' education programmes.
- Implementation of an aided self-help programme, which provided technical and material resources from the government, matched with people's labour and food to enhance the basic infrastructure of these communities.
- Establishment of community groups and cooperatives in communities to enable them to enhance sustainable livelihoods, good governance, respect for the law, peace and partnership in poor, marginalized communities. For example, the Fancy Farmers' Cooperative bought manure and farm supplies in bulk at a cheap rate and then sold them to farmers at cheaper/discounted prices. They encouraged cooperative farmers in the rural communities to grow certain cash crops and then repurchase them from these farmers at a set price which helped to sustain the livelihoods of the people in remote rural villages like Fancy.
- Participatory mapping and cultural wakes are considered essential development practices as they showcase the vulnerability of climate change and other development calamities in the community. These strategies empowered communities to collectively develop strategies to resolve the development problems facing these communities.
- The CDD programme reveals a hybrid approach where communities work collectively with governmental agencies and other external partners to implement basic needs projects. CDD shifted in direction resulting in the facilitation of development programmes at the community level by brokering funding agencies to access financial and technical resources to resolve community-felt needs. It also advocates with the central government for equitable distribution of resources to communities.

3.5.9 Recommendations

The CDD is a government agency that initially implemented a top-down approach to addressing community development challenges. However, due to the failure of this approach, a hybrid development approach that combines the strengths of top-down and bottom-up approaches was implemented (Matland, 1995; OECD, 2013; Butler et al., 2015). The approach combines participatory approaches to empower people to lead and shape the type of development they desire to implement in their respective communities. The role of government transformed from an implementor of development to a catalyst and facilitator of the development process. The approach enhanced the levels and quality of people's participation in community development (Chambers, 2002; Simmons, 2021).

In 1997, the director of community services left the Ministry of Social Development to take up an assignment at the Commonwealth Secretariat in London, UK (Searchlight Newspaper, 1997). One year later, the CDD project began to lose momentum due to a change in government policy and approach back to a top-down development approach that reduced the participation of the community in the development process. This situation was compounded when DfID terminated funding of the CDD programmes in St. Vincent and the Grenadines. The situation worsened when there was a government change after the general election a few years later, resulting in the community development strategy no longer being a priority of the Government of St. Vincent and the Grenadines.

The recommendations to sustain the momentum of the CDD programmes are as follows:

- There was evidence of mistrust between politicians and the communities they serve, resulting in politicians taking financial resources from governmental agencies such as CDD and beginning to spend the resources directly in the communities. In most circumstances, these projects are not based on the community's felt needs, and neither are the people given any opportunity to participate in the identification and implementation of such projects. As were the practices of CDD to address such situations from recurring, the CDOs should work with the parliamentary representatives and the community leaders to build back trust and partnership among both stakeholders.
- The CDD should continue its effort to inject funding and technical resources to develop education and training programmes to enhance the awareness and overall capacity of the community to develop projects and programmes to resolve the development challenges affecting these communities.
- Unemployment is viewed as a critical developing challenge affecting communities in St. Vincent and the Grenadines, mainly affecting youth and women. People must be gainfully and reasonably employed in decent jobs to prepare for themselves and their families. The government should reassess the skills provided by technical and tertiary institutions when in the context of opportunities available now and in the future and retail the curriculum and education system to best respond to the employment needs. The new emphasis should include training youth and women in entrepreneurial skills to enable them to set up their own businesses utilizing technologies as a vehicle for marketing their products.
- The government needs to invest in developing sustainable livelihood programmes linked to the country's natural resource base, including agriculture, ecotourism and fisheries. The government should explore investing in projects to increase the production of the arrowroot and fisheries by exploring initiatives to add value to these products by exploring opportunities around canning and manufacturing other products.
- Limit the political interference of the parliamentary representative and their agents in the programmes of the CDD by allowing the community's staff and people to develop and implement the projects.
- The injection of financial and technical resources in the CDD to enhance the leadership capacity of community leaders by enhancing their conflict resolution, climate change adaptation, mobilization, communication and management skills and knowledge.
- Continue work on the building of partnerships and cooperation with internal and external agencies to mobilize financial resources to build the capacity of the staff of CDD to implement viable and sustainable programmes and projects at community levels.

3.6 Summary of Outcomes of the Chapter

One of the essential lessons evolving from the case studies on JEMS and the Community Development Division, Government of St.

Vincent and the Grenadines, is that community participation in consultative participatory development processes is essential to the success of sustainable development programmes/projects (Hernandez-Delgado, 2015; Jaja et al., 2016). The CDD top-down approach to the development of community development programming failed miserably in St. Vincent and the Grenadines, while comparatively, JEMS was implementing a successful bottom-up approach to the development of communities while the government's community development programmes were failing.

In 1990, the government employed new leadership in the Ministry of Social Development and shifted its approach to a hybrid approach. The hybrid approach combines the strengths of top-down and bottom-up approaches engaging people in a participatory process geared towards empowering them to develop the communities across St. Vincent and the Grenadines. The new approach provides the division with a renewed momentum and a glimmer of hope resulting from the implementation of successful projects (Matland, 1995; OECD, 2013; Stachowiak et al., 2016). The new leadership was accredited for the success of the JEMS initiative and was trained at the tertiary level, skilled in applying community development models, and possesses a wealth of experience implementing community development strategies in local and regional communities in 16 countries across the Caribbean (Rittgens & La Gra, 1991; Leach, 1996).

The CDD implemented an intense training programme to enhance the staff's knowledge and skills in innovative participatory community development practices. The division employed the services of veteran gender and community development practitioners such as Chaddis Stapleton and Amour Jackson to work as CDO in the IVD programme. Former executive members of the National Youth Council (NYC) David Latchman and Sullivan Harry were employed as CDOs to engage the youth in the community development programme. Other strategies employed include (a) mobilization of scholarships from the DfID for staff to pursue studies overseas; (b) implementation of quarterly training workshops for staff on various aspects of development; (c) organizing of internships, exchanges and study tours both nationally and regionally to provide staff and community leaders with first-hand experiences on the implementation of sustainable community development practices; and (d) provision of tuition fees and other educational provisions to support staff who lack the minimal requirement for full-time employment in the civil service. Hon. Louis Jones, minister of social development, fully supported these capacity-building initiatives.

The director of community services continues his voluntary role as president of JEMS while working in the government's public service. The JEMS organization provides an opportunity for piloting innovative CD models and approaches before implementing them in the government CDD's programmes. The JEMS provides a wealth of data on what strategies work and that would not work. These innovative strategic programmes were tested 6–9 months before they were integrated into the national CD programmes. The CD officers were involved in joint capacity-building training programmes with JEMS in project writing, project management and implementation, community mobilization and learning, leadership development and conflict resolution. This provides exposure to the CDOs and facilitates the transfer of skills and knowledge on strategies for implementing sustainable development programmes at a community level.

JEMS had its own challenges in implementing community participatory development actions in the early stages of the organization. It took years of training and programming before the membership of JEMS could implement successful community development initiatives. The organization developed its participatory development models, a clear vision and a committed and dedicated membership after years of "trial and error" in implementing community sustainable development programmes. "We started going into the community, telling people to stop cutting down the trees. But I think we learned a harsh lesson when we were confronted by one parent who told us…how do you expect me to get food? I can't buy natural gas; how do you expect me to get food to feed my children, to send them to school?"...So, we started to have discussions. We changed our entire strategy in terms of how people within the community could work with us to save the forest. One of the major problems in that community was illiteracy. Illiteracy was estimated at over 60% among adults. So, we started a literacy program. We ran classes four evenings per week. I recruited other teachers and other persons in the organisation who were going to secondary school. They worked as volunteers. I trained them to work with others", said Simmons, founder and former president of JEMS (Leach, 1996, p. 1).

Before 1990, there was a breakdown in trust between communities and the officials at the government's CDD. The financial resources allocated to the CDD were mainly utilized in one constituency. Community and non-governmental organizations such as the National Youth Council were experiencing difficulties accessing resources from the CDD. The community leaders accused the CDD project approval process of lacking transparency. Therefore, the employment of the new director of community services resulted in the creation of a dynamic and coherent sustainable development model, which contributed to enhancing trust and transparency in the CDD process.

Although the government provided matching funds to support the implementation of the CDD programmes, a larger percentage of the funds mobilized to support the programmes of both organizations was through building partnerships with local, national, regional and international development agencies. The partnership of CDD with the DfID, OAS, USAID and others was documented earlier in the case study. Through the partnership, JEMS received funding from the UNDP GEF SGP to implement environmental programmes at the community and national levels. The JEMS/UNDP/GEF projects increased community understanding of the Global Environment Facility's thematic areas. The outcomes from the participatory community activities resulted in setting up a National Project Appraisal Committee (NPAC) for prioritizing and pre-screening all potential GEF SGP projects in St. Vincent before sending them to Barbados for approval and payments. This National Project Appraisal Committee functions through the Ministry of Environment. It was approved by the Government of St. Vincent and the Grenadines and the UNDP. JEMS held the committee's deputy chair for a considerable number of years.

Despite the development challenges associated with JEMS regarding the stern position held by the organization regarding the Government of St. Vincent and the Grenadines implementing the Argyle International Airport and the cross-country road without an approved environmental impact assessment (EIA) study, the organization will always be ready to work with the government to implement its national development programme (https://sgp.undp.org/spacial-itemid-projects-landing-page/spacial-itemid-project-search-results/spacial-itemid-project-detailpage?view=projectde); JEMS was also requested by the Government of St. Vincent and the Grenadines to sit on the Project Supervisory/Appraisal Committee during the construction of the Argyle International Airport.

References

Angelsen, A., Larsen, H. O., Lund, J. F., Smith-Hall, C., & Wunder, S. (2011). *Measuring livelihoods and environmental dependence: Methods for research and fieldwork*. Earthscan.

Butler, J. R. A., Wise, R. M., Skewes, T. D., Bohensky, E. L., Peterson, N., Suadnya, W., Yanuartati, Y., Handayani, T., Habibi, P., Puspadi, K., Bou, N., Vaghelo, D., & Rochester, W. (2015). Integrating top-down and bottom-up adaptation planning to build adaptive capacity: A structured learning approach. *Coastal Management, 43*, 346–364.; Taylor & Francis Group, LLC; ISSN: 0892-0753 print /1521-0421 online. https://doi.org/10.1080/08920753.2015.1046802

CARIBSAVE. (2012). *Climate change Risk profile for St. Vincent and the Grenadines*. CARIBSAVE.

Chambers, R. (2002). *Participatory workshops: A sourcebook of 21 sets of ideas and activities* (Vol. 794, pp. 421–434). Earthscan.

Commodore, A., Wilson, S., Muhammad, O., Svendsen, E., & Pearce, J. (2017). Community-based participatory research for the study of air pollution: A review of motivations, approaches, and outcomes. *Environmental Monitoring and Assessment, 189*(art. 378).

Danielsen, F., et al. (2009). Local participation in natural resource monitoring: A characterization of approaches. *Conservation Biology, 23*, 31–42.

Eicken, H., Danielsen, F., Sam, J. M., Fidel, M., Johnson, N., Poulsen, M. K., Lee, O. A., Spellman, K. V., Iversen, L., Pulsifer, P., & Enghoff, M. (2021). Connecting top-down and bottom-Up approaches in environmental Observing, *BioScience, 71*(5) 467–483.

Enemark, S. (2003). Understanding the concept of capacity-building and the nature of land administration systems. *FIG Working Week*, April, Paris, France.

Fox, L. M., & Schearer, S. B. (1997). *Sustaining civil society: strategies for resource mobilization*. Washington, D.C., CIVICUS.

Fox, N. S., Brennan, J. S., & Chasen, S. T. (2008). Clinical estimation of fetal weight and the Hawthorne effect. *European Journal of Obstetrics, Gynecology and Reproductive Biology, 141*(2), 111–114.

Gomez, O. (2013). *Climate change and migration: A review of the literature*. International Institute of the Social Studies, The Hague (Erasmus University Rotterdam).

Government of St. Vincent and the Grenadines. (2008). *St Vincent and the Grenadines country poverty assessment 2007/2008L living conditions in the Caribbean in a Caribbean Small Island Developing State Volume 1 Living Conditions in St. Vincent and the Grenadines*. Kingstown.

Government of St. Vincent and the Grenadines. (2013a). *National economic and social development plan 2013–2015: Re-engineering economic growth: Improving the quality of life, Kingstown, SVG.*

Government of St. Vincent and the Grenadines. (2013b, July). *National Report St. Vincent and the Grenadines: Third international conference on small island developing states – National report: Ministry of Health Wellness and the Environment.*

Grove, R. H. (2000, April 18). *The culture of Islands and the history of environmental concern* (pp. 2–10). Harvard Seminar on Environmental Values, Harvard Divinity School, Harvard University. Available at: http://ecoethics.net/hsev/200004txt.htm. Accessed on 26 Dec 2017.

Hernandez-Delgado, E. A. (2015). *The emerging threats of climate change on tropical coastal ecosystem service, public health, local economies, and livelihoods sustainability of small islands: Cumulative Impacts and Synergies.* Marine Pollution Bulletin, University of Puerto Rico, Centre for Applied Tropical Ecology and Conservation, Coral Reef Research Group Puerto Rico, USA.

https://data.worldbank.org/indicator/SP.POP.DPND.YG?locations=VC

https://pdf.usaid.gov/pdf_docs/PNADE352.pdf

https://sgp.undp.org/spacial-itemid-projects-landing-page/spacial-itemid-project-search-results/spacial-itemid-project-detailpage?view=projectde

https://warontherocks.com/2019/02/in-defense-of-best-practices/

https://www.goldmanprize.org/recipient/andrew-simmons/

ILO (2010) *A skilled workforce for sustainable growth: A G20 Training Strategy*. International Labour Office, Geneva, November 2010.

IMF. (2016). *St. Vincent and the Grenadines: Staff Report 2016 Article 1V consultation*. International Monetary Fund.

IMF. (2017). *St. Vincent and the Grenadines, IMF Country Report NO 17/400, Article IV Consultation with St. Vincent and the Grenadines*. International Monetary Fund, Press Release No. 17/400, Washington, USA, December 2017.

IPCC (2014a) *Synthesis Report. In: Contribution of Working Groups I, II and III to the Fifth Assessment Report of the Intergovernmental Panel on Climate change*, IPCC Geneva, Switzerland, pp.151.

IPCC (2014b) *Summary for Policymakers. In: Climate change Impacts, Adaptation and Vulnerability*. In: C. B., Barros, V. R., Dokken, D., Mach, K. J., Mastrandrea, M. D., Bilir, T. E., Chatterjee, M., Ebi, K. L., Estrada (Eds.) Contributions of the Working Group II to the Fifth Assessment Report, Field. IPCC Geneva, Switzerland. https://www.ipcc.ch/assessment-report/ar6/

IPCC (2014c) Climate change: Synthesis Report. In: Contribution of Working Group 1, 11, 111 to the Fifth Assessment Report of the Intergovernmental Panel on Climate change, IPCC Geneva, Switzerland pg.151.

IPCC. (2022). *Climate change 2022 mitigation of climate change, summary for policy makers: Working group 111 contribution to the sixth assessment report of the international panel on climate change (AR6 WG 111)*. IPCC

Jaja, J., Dawson, J., & Gaudet, J. (2016). *Using social network analysis to examine the role that institutional integration plays in community-based adaptive capacity to climate change in Caribbean small island communities*. Department of Geography, Environment, and Geomatic, University of Ottawa, Canada: Routledge, Local Environment.

Khan, M. R., Timmons Roberts, J., Huq, S., & Hoffmeister, V. (2018). *The Paris framework for climate change capacity-building*. Routledge; Taylor and Francis Group; Earthscan.

Lazrus, H. (2012). Sea change: Island communities and climate change. *Annual Review of Anthropology, 41*, 285–301.

Leach, M. (1996). *A revolution of ideas and action: Interview with Andrew Simmons: Share International: July/August 1998 Issues of Share International*. https://www.share-international.org/archieves/environmental/en_mlrevolution.htm (download 09/05/2022).

Matland, R. (1995). Synthesising the implementation literature: The ambiguity-conflict model of policy implementation. *Journal of Public Administration Research and Theory, 5*(2), 145–174.

OECD. (2013). *The nature of policy change and implementation: A review of different theoretical approaches*. Organisation for Economic Cooperation and Development.

Poschen, P. (2015) *Decent work, green jobs and the sustainable economy: Solution for climate change development*. International Labour Organisation, Geneva, Switzerland.

Rittgens, R. & La Gra, J. (1991). *Organisations in St Vincent and the Grenadines: IICA: IICA Office in St Vincent and the Grenadines, Kingstown, St Vincent and the Grenadines*.

Robson, C. (2002). *Real world research, a resource for social scientists and practitioner researchers*. Blackwell Publishers.

Saunders, M., Lewis, P., & Thornhill, A. (2000). *Research methods for business* (2nd ed.). Pearson Education Limited.

Searchlight Newspaper. (1997, October 3). Simmons Bound for London. *Searchlight Newspaper*, Friday, *3*(40).

Simmons, A. (2018). *Impact of climate change on youth in small island communities: The Case of St. Vincent and the Grenadines*. This thesis is submitted to De Montfort University in partial fulfilment of the requirements for the award of Doctor of Philosophy: School of Engineering and Sustainable Development De Montfort University, Leicester, UK

Simmons, A. (2021). *Impact of climate change on youth in small Island developing communities* (Sustainable development series). Springer.

Simmons, A. (2022). *Climate change adaptation framework and youth entrepreneurship in West Africa* (Sustainable development goals series). Springer.

Stachowiak, S., Robles, L., Habtermanion, E., & Maltry, M. (2016). *Beyond win: Pathway for policy implementation*. ORS Impact/The Atlas Learning Project.

Third World Resurgence. (1992). No 24/25 1991, P 27.

UNDP, (2015). *Human Development Report; Work for Human Development*, United National Development Programme (UNDP) New York, USA.

Yin, R. K. (2014). *Case study research: Design and methods*. Sage.

Capacity-Building of Institutions for Climate Change Adaptation and Sustainable Development in a Post-conflict LDC

4

Abstract

This chapter presents case studies on capacity-building in Sierra Leone, West Africa. The first section of the chapter focuses on implementing the capacity-building initiative through the Youth Employment and Empowerment Programme (YEEP) in a post-conflict environment by the Government of Sierra Leone in collaboration with the UNDP SL. The project focuses on strengthening the capacity of the National Youth Commission (NAYCOM) and the Ministry of Youth Affairs (MoYA), the implementation of the "Youth Development Partner Coordination Framework", the development of a blueprint of the youth development programmes 2014–2018 and the mapping of donor support to youth development projects in Sierra Leone as well as providing training in skills which leads to employment creation and sustainable livelihoods.

The chapter presents an analysis of the top-down strategies and outcomes of the UNDP-funded Youth Employment and Empowerment Programmes (YEEP) in Sierra Leone, highlighting the gaps, challenges, strategic approaches, best practices and lessons learned regarding building the capacity of youth and development institutions to resolve youth employment and poverty challenges. The second part of the chapter features a bottom-up approach implemented by the Global Youth Network for Empowerment and Development (GYNED) in building the capacity of youth in Sierra Leone through entrepreneurship and sustainable livelihoods.

Keywords

Sierra Leone · International development · Youth employment · Entrepreneurship · West Africa · Sustainable livelihood · Poverty · Capacity-building · Top-down approach · Bottom-up approach · Hybrid approach

This chapter presents case studies on capacity-building in Sierra Leone, West Africa. The first section of the chapter focuses on implementing a capacity-building initiative through the UNDP-funded Youth Employment and Empowerment Project (YEEP) in a post-conflict environment by the Government of Sierra Leone in collaboration with the UNDP SL. The project focuses on (a) strengthening the capacity of the National Youth Commission (NAYCOM) and the Ministry of Youth Affairs (MOY), (b) the implementation of the Youth Development Partner Coordination Framework, (c) the development of a blueprint for the youth development programmes 2014–2018 and (d) the mapping of donor support to youth development programmes in Sierra Leone

A. Simmons, A.-A. Simmons, *Building Institutions for Resilience*, Sustainable Development Goals Series, https://doi.org/10.1007/978-3-031-28881-4_4

and the provision of training in employment and entrepreneurship skills to youth which leads to sustainable livelihoods. The chapter features a bottom-up approach implemented by the Global Youth Network for Empowerment and Development (GYNED) to build the capacity of youth in Sierra Leone through the development of entrepreneurship and sustainable livelihood programmes. The chapter then presents an analysis of the top-down strategies and outcomes of the UNDP-funded Youth Employment and Empowerment Programmes (YEEP) in Sierra Leone, highlighting the gaps, challenges, strategic approaches, best practices and lessons learned regarding building the capacity of youth and development institutions to resolve youth employment and poverty challenges.

4.1 Global Youth Network for Empowerment and Development Sierra Leone (GYNED-SL): Bottom-Up Approach to Youth Development and Sustainability Livelihoods in Sierra Leone, West Africa

Sierra Leone is a least developing state (LDS) located in West Africa. It has a population of 7 million, of which 1.7 million are categorized as young. Approximately 75% of its population is below the age of 35. It is estimated that over 60% of young people are unemployed or underemployed. These young people are unable to provide sufficiently for themselves and their families. The severe unemployment and poverty facing young people in Sierra Leone are compounded mainly by failing secondary and tertiary education systems. Approximately 37% of school-age youth are currently in education, and half of all young people are defined as functionally illiterate, i.e. they do not possess the minimum literacy and numeracy skills to function effectively in modern society.

According to the OECD, literacy is "the ability to understand, evaluate, use, and engage with written texts to participate in society, achieve one's goals, and develop one's knowledge and potential" (OECD, 2013, p. 59). A person is functionally illiterate when they cannot engage in all the activities in which literacy is required for the effective functioning of their personal and the community development needs (UNESCO, 1978; Vagvolgyi Raka et al., 2016). The situation in Sierra Leone and other LDCs is dire and places huge stresses and strains on social protection systems regarding the demand for services and the opportunities lost in economic production and revenue generation. It also collides young people with the country's social and political life. Service sector businesses, public sector reform and growth in the extractive industries are all at a halt. However, the agriculture and fisheries sectors offer the greatest potential to generate the hundreds of thousands of jobs needed to build resilience and buoyancy in the economy of Sierra Leone. Despite positive signs of growth in these sectors, climate change continues to be the single factor which is devastating the natural resources bases of Sierra Leone (IPCC, 2014, 2022; Simmons, 2018, 2021, 2022). Therefore, achieving any form of resilience among the people and their communities would require a significant injection in human capital and technological resources supported by policy reform, political will and practical actions on the part of the Government of Sierra Leone.

Despite the challenges highlighted above, there are signs that the civil society movement in Sierra Leone is emerging from a bottom-up perspective as an effective pillar of the development process. However, the challenges facing the youth sector are based on the ability of the sector to develop the consensus, organization and influence to effectively drive this change in collaboration with young people. The Global Youth Network for Empowerment and Development Sierra Leone (GYNED-SL) is one of the premiere bottom-up youth development organizations in Sierra Leone that emerged in the process of driving such development forward. The section of the book will feature the work of the GYNED-SL, focusing on its strengths and weaknesses, the effectiveness of the methodologies implemented, lessons learned and best practices evolving from the organization's capacity-building programme.

4.1.1 Global Youth Network for Empowerment and Development Sierra Leone (GYNED-SL)

GYNED-SL is a civil society organization based in Sierra Leone established in 2012. Civil society organization is defined as the third sector of the economy and includes a wide array of relief and welfare organizations, innovation organizations, public service organizations, economic development organizations, grassroots organizations and mobilizing groups. The sector plays a critical role in creating values worldwide through its work in the delivery of services, advocacy, cultural programmes and social movements. The organization was created to deal with the inhumane and poverty-stricken conditions confronting young people in Sierra Leone. Despite the positive efforts of the governments working in partnership with the international community to invest in sustainable development initiatives, the severity of the development challenges continues to cause uncertainties among the young people in Sierra Leone.

The organization's leadership possesses the enthusiasm and drive to resolve the development challenges evolving from decades of implementing welfare-oriented programmes in Sierra Leone. This situation is compounded by the impact of a decade-long civil war and, in more recent times, the impact of the Ebola epidemic on the lives and livelihoods of the Sierra Leonean society. The organization's overall purpose is to transform and empower Sierra Leonean young people, especially young women and girls, from the apprehension and periphery of the society in Sierra Leone. The organization is legally registered as a non-governmental organization (NGO) in the annals of the Sierra Leone Association of Non-Government Organizations (SLANGO); the Ministry of Planning and Economic Development (MoPED); the Ministry of Social Welfare, Gender and Children's Affairs (MSWGCA); the Ministry of Youth Affairs (MoYA); the Ministry of Education (MoE); and the National Youth Commission (NAYCOM). It is a member and NGO representative on the Steering Committee of the TVET Coalition of Sierra Leone, the Human Rights Working Group and the Salone Adolescent Girls Network.

The GYNED-SL is a national network organization with operations throughout the 16 districts of Sierra Leone. Its membership comprised youth and community representatives, volunteers and other member organizations. The organization targets young women, girls and other minority groups due to the many development challenges encompassing their growth and development, particularly in LDCs such as Sierra Leone. Young women and men continue to face deprivation, violence and other forms of abuse daily and are often the perpetrators and victims of violent crime (Simmons, 2006; Charles, 2006).

4.1.2 GYNED Capacity Development Building Initiatives

GYNED-SL has 22 full-time staff, 30 district representatives, 48 member organizations and over 3000 individual youth volunteers across all the districts. These representatives are responsible for developing and implementing bottom-up innovative youth sustainable development programmes that respond to the community's needs (Matland, 1995; Danielsen et al., 2009; Commodore et al., 2017; Eicken et al., 2021). The organization's work centres on implementing innovative projects and programmes utilizing the latest project management technologies and strategies. It produced accurate and timely reports and results that sustainably impacted youth and their communities. GYNED-SL is a change management-oriented organization that practices the four steps in driving positive change processes across Sierra Leone at the chieftain, district and national levels. The organization possesses strong leadership and a positive change culture (Burnes, 2000; Johnson & Scholes, 2002). GYNED-SL utilizes a participatory bottom-up approach to initiate, plan and implement its projects and programmes in relation to the context and dynamics of the beneficiaries and communities. The organization implements rights-Based and needs-based approaches to development at district and national levels (Matland, 1995; Chambers, 2002; Commonwealth Secretariat, 2007; Simmons, 2018, 2021).

GYNED-SL implemented several projects with the National Youth Commission, YMCA, Lattanzio and others with funding from the UNDP, GIZ, Comic Relief, African Development Bank (AfDB) and other institutions. Musa Koroma, executive director, states that the organization created its own model for driving its work with young people forward and engaging them to participate in its development programmes, including (a) connect, (b) inspire, (c) motivate and (d) change (Fig. 4.1).

GYNED-SL views each of the above-stated elements as an integral component in achieving a positive, strong and vibrant youth actively involved in leading the national sustainable development process, thus making youth voices heard across the country. GYNED-SL would collaborate with a range of community, district and national development organizations to ensure that it effectively connects with Sierra Leonean youth. Such partnerships enable the leadership of GYNED-SL to work mutually by sharing responsibilities, resources and ideas with key stakeholders on a national stage by engaging and inspiring young people and key stakeholders to become socially conscious and responsible citizens (Feiner, 2002; Becker & Kretsch, 2019). The organization has inspired and motivated individuals to give their best to achieve the desired result, i.e. it is about getting people to move in the right direction, gaining their commitment and motivating them to achieve their desired goals (Armstrong, 1999):

> A motivated conscious, and skilled young people in Sierra Leone can ignite positive change, and the organisation stands ready to guide these young leaders who are on their journey to achieving sustainable development goals. Musa Koroma, executive director, GYNED-SL

4.1.3 The Vision and Mission of GYNED-SL

The organization's vision is to "Envision a poverty-free world where young people can maximise their fullest potentials" (GYNED-SL, 2020, p. 7).

The mission of GYNED-SL is to "Build capacities and advocate for the improvement of the socio-economic status of the marginalised and less fortunate people and promote the social rights of young people through collective sensitisation and training; seeking to positively influence and empower them to develop their self-esteem and confidence and to become self-reliant instilling in them positive leadership values and qualities in exploring their future possibilities" (GYNED-SL, 2020, p. 8).

4.1.4 Programmes Implemented

GYNED-SL is a dynamic youth development organization based in Freetown, Sierra Leone, which advocates on behalf of youth and delivers services targeted at the holistic development of young people, women, girls and their communities. The programmes implemented by the organization include:

(a) *National Youth Development Summit*

The Youth Development Summit aimed at creating opportunities for elevating the voices of

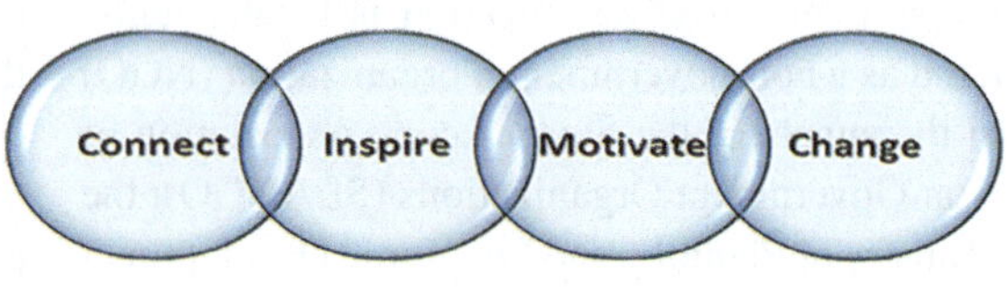

Fig. 4.1 GYNED-SL model for working with youth

young people and placing their experiences and concerns at the centre of decision-making, as well as providing them with prospects for engaging with a range of development partners on the development priorities of the country. The Summit also highlights progress made by the Government of Sierra Leone on implementing the National Youth Policy and the operationalization of the Africa Union Roadmap on Harnessing the Demographic Dividend through Investments in Youth. These initiatives were assessed in line with the Government of Sierra Leone's Medium-Term National Development Plan (2019–2023), the UN Sustainable Development Cooperation Framework and the UN Youth 2030 Strategy. The National Youth Development Summit was designed as a platform to launch the National Youth Economic Empowerment Fund (NYEEF) in collaboration with the Ministry of Youth Affairs. The aim of the fund is to work towards enhancing the sustainability and resilience of youth towards the achievement of the Sustainable Development Goals (Government of Sierra Leone, 2013).

(b) *Climate Change Action and Environmental Protection and Education*

The Climate Change Action and Environmental Protection initiative intend to develop and strengthen the institutional capacity of youth and community-based organizations (CBOs) by establishing environmental protection clubs (EPC) in primary and secondary schools within the Tengbeh Town Community, implementing village clean-up actions and training community leaders in climate change adaptation practices (GYNED-SL, 2020). The leadership of the participating CBOs was trained to develop strategies to resolve land degradation, solid and liquid waste dumping issues, deforestation and other environmental challenges at community levels. GYNED-SL was responsible for implementing training, coordination and the project management aspect of the initiative. The climate change and environmental programmes were implemented in collaboration with a committee of experts in environmental protection. GYNED-SL worked towards establishing itself as a centre of excellence and learning in community climate change adaptation, environment and sustainable management. The organization facilitated the development, learning and exchange of best practices in community climate change adaptation and sustainable development initiatives at community and district levels (GYNED-SL, 2020).

The organization worked with CBOs, women groups, households, primary and secondary schools, unemployed youths and a range of key stakeholders who are directly affected by the challenges of climate change, land degradation and the impact of broader environmental issues. GYNED-SL played advocacy, catalytic and brokerage roles by bringing the vulnerable and marginalized people who are impacted by climate change in direct contact with officials from the Ministry of Environment, the Environmental Protection Agency and other governmental agencies to develop a partnership to (a) assess the consequences of their actions and (b) devise practical strategies for addressing their problems (Angelsen et al., 2011; Nurse et al., 2014; Hernandez-Delgado, 2015).

(c) *Youth Entrepreneurship Development and Growth*

Unemployment and poverty are severe development challenges affecting youth in Sierra Leone and other developing countries in West Africa (ILO, 2010; UN, 2015; Simmons, 2022). GYNED-SL possesses unique skills and competencies in utilizing entrepreneurship to resolve youth unemployment and underemployment needs. The organization collaborated with the National Youth Commission of Sierra Leone to implement the Youth Employment and Empowerment Project (YEEP). The overall purpose of the YEEP was to scale up the ongoing implementation of the project in two components, namely, (a) Skills Development for Employment and (b) the Graduate Entrepreneurship Programme (GEP). The specific objectives of the project were (a) to improve the efficiency and productivity of young entrepreneurs by creating enhanced interactive learning platforms through training in entrepreneurship and business development, fish farming management and technical skills and (b) to revitalize the entrepreneurial spirit of the youths through the establishment of business incubation and growth programmes that empowers them to develop effective business innovations and entrepreneurial ecosystems. The project has a gender perspective focusing on empowering young women and girls by enhancing sustainable livelihoods through skills training and entrepreneurship development. It also utilized a gender mainstreaming approach to mainstream gender issues across institutions at the national, district and chiefdom levels. The African Development Bank (AfDB) and the UNDP funded the project (ILO, 2010; UNDP, 2015; Simmons, 2021, 2022).

(d) *Education and Capacity Development*

Education and training of youth is an important component of the GYNED-SL strategic programme. To develop the case study, the author highlighted an innovative, short-term low-value grant initiative that supports 300 COVID-19-affected youths living in slums and on the street within Freetown, the capital of Sierra Leone. The programmes delivered social services, including psychosocial support and life skills. The primary purpose of the Education and Capacity Development project was to restore and improve the well-being of youth slum dwellers on a mental/psychological and social/collective level. Other education and capacity-building programmes implemented by GYNED-SL include youth and ICT, WASH education and management actions, entrepreneurial development, TVET and agribusiness development programmes (Chanita-Mataya et al., 2020).

4.1.5 Strengths of GYNED-SL Capacity-Building Initiatives

The respondents who participated in the focus group discussion (FGD) categorized the strength of the GYNED-SL as follows:

> The organisation has good human resources which are very good at implementing projects at chieftain, district, and national levels. It has excellent leadership, which makes the organisation unique and outstanding. We have a cadre of qualified staff who are trained and certified at tertiary level in appropriate development disciplines. Alvian B. K. Sesay, communication officer of GYNED-SL

This view is supported by Musa Koroma, founder and executive director of GYNED-SL, as he posits:

> We ensure that the projects are delivered to the persons who need them. The communities are different, so you involve the chief, the community leaders, and the beneficiaries on the ground where the projects are to be implemented to make decisions on these projects. After this consultation, we inform the donors about the project on the ground...Involving people directly or indirectly as beneficiaries is critical to our participatory project implementation process. The use of diplomacy when working with communities to resolve problems is also of critical importance...It is also important to involve the older leaders in the development process.

GYNED-SL is a unique youth sustainable development organization that possesses distinctive strengths. The organization is registered and recognized in Sierra Leone. It is recognized by the Government of Sierra Leone and other leading youth development organizations, including the government's National Youth Commission and the Ministry of Youth and international development organizations, including bi- and multilateral organizations, IGOs and other NGOs. It is staffed by a team of young professionals and volunteers who are passionate, dedicated and committed to working with youth. The scope of the organization's work is nationwide and driven by an army of volunteers working in district and community levels. The agency possesses a unique bottom-up participatory approach to project management and has successfully implemented a range of youth development projects in collaboration with MoYA, NAYCOM, UNDP, UNFPA, AfDB, GIZ and others. These projects concentrate on building the resilience of youth and their communities to adapt to climate change's impact and addressing a range of other development calamities (Matland, 1995; Butler et al., 2015).

The organization helps young people become volunteers within their community and district levels by building their capacity through training, internship, mentoring and coaching and exposing them to participate in the implementation of educational and livelihood projects in their communities (Chania-Mataya et al., 2020). It trained a cadre of youth volunteers who participated in various development projects across the country.

4.1.6 Weaknesses of the GYNED-SL Development Model

Although GYNED-SL possesses unique capacity development and organizational strength, its weaknesses continue to hinder the organization from achieving its development goals:

> There is a culture in Sierra Leone where NGOs are not being provided with resources to sustain themselves. The strategy is to work as partners with agencies such as UNDP and GIZ to mobilise much need resources to implement development projects. The situation is compounded by lack of skills and knowledge within NGO and civil society sector to write projects on employment, sustainable livelihoods and climate change issues. Musa Koroma, executive director of GYNED-SL

GYNED-SL, like most NGOs operating in LDCs, possesses its total share of weaknesses. Despite being known to the government and the international community for implementing successful youth development projects in remote and marginalized areas in Sierra Leone, it is finding it extremely difficult to mobilize adequate financial and material resources to support the implementation of its programmes at district and community levels. The agency has an extensive portfolio of unfunded projects submitted on behalf of the residents of marginalized urban and rural communities. The lack of funding to support such needed initiatives has caused undue frustrations and stress on these communities and the organization.

The organization lacks key and experienced experts among the staff skilled in pertinent areas of sustainable development. Due to financial constraints, the organization has difficulties hiring and maintaining experienced staff to implement its programmes. It heavily relies on volunteers and interns from universities and colleges to sustain its project implementation momentum in the field. The lack of available funding also puts undue pressure on the organization to purchase new equipment for its job readiness and skills training programme. Although the organization recently acquired adequate space to run its programmes, a considerable amount of time and resources are expended to mobilize resources to cover the office and training space rent. Therefore, time and resources would have been better spent by staff on working with communities to develop and implement its programmes rather than expending most of its energies on fundraising.

4.1.7 Major Challenges and Strategies for Resolving Such Development Challenges

The organization's major challenges were classified as (a) financial and (b) capacity development:

> Lack of financial and material resources to support project programmes in the field is viewed as being very serious...the organisation needs funding to support the implementation of the long-term project. Moses Kalokoh Putin, media official, GYNED-SL

The civil society and NGO sector in Sierra Leone do not work in close partnership regarding developing the country:

> There is too much competition from NGOs in Sierra Leone for the small pot of money available to NGOs to implement projects. There are lots of good organisations with good projects not getting the funding needed to implement their projects. The donor agencies should give funding to those agencies who are working in communities to implement their projects. Donor work with NGOs to put stringent measures in place to ensure the target beneficiaries and their communities benefit from the project. Aminata Olimatu Kamara, a staff member of GYNED

The major development challenge facing GYNED-SL is its inability to mobilize adequate resources to support the organization's long-term programme development, implementation and sustainability. Such resources are categorized as (a) finances to fund the capacity development programmes implemented by the organization at national and district levels; (b) provision of tools such as monitoring, evaluation and reporting (MER) framework and appropriate technologies to support the effective implementation of training, networking, internship and mentoring programmes; and (c) availability of professional staff and experts to assist with the development of projects, mobilization of resources and implementation of training and other capacity development programmes in-house and at the national level.

The organization continues to explore opportunities for resource mobilization by providing volunteers to support the activities of MoYA and NAYCOM and other governmental agencies. The strategy is based on the notion that if the GYNED-SL offers services to these governmental agencies without any payment for such cost, these agencies will subcontract the work to GYNED-SL when they are provided with government and donor funds. Musa Koroma, executive director, states that the strategy returns dividends, as seen in the implementation of the YEEP with NAYCOM and the AfDB. GYNED-SL also continues to explore opportunities to strengthen local and international partnerships as a strategy to resolve the capacity development challenges confronting the organization.

4.1.8 Capacity Development Problems (Human) Facing the Organization

GYNED-SL, like most youth NGOs, has its share of capacity development problems. It lacks proficiency in youth development and project management:

> The team lacks the key expertise in project management and delivery. It also needs experts who understand the issues relating to youth development and sustainable development. The experts required should be endowed with skills on strategies to enhance the organisations' skills development programme. Musa Koroma, executive director of GYNED-SL

In recognition of the deficiency in capacity, the organization is implementing a series of workshops to enhance the staff's capacity as articulated by Ostilda C. O. Hughes, operations director, staff member:

> We have workshops every Monday to provide the opportunity for the sharing of skills among the members of the organisation.

The main capacity development problems were identified by staff and volunteers as the lack of key experts with expertise in project management and resource mobilization to support the organization's programmes. The GYNED-SL needs experts who are skilled in the development and implementation of projects as well as the mobilization of financial resources to support the implementation of projects at the district, chiefdom and community levels. Due to the severity of the impact of climate change on the productive sector of Sierra Leone and the complexities regarding the development of climate change adaptation projects, GYNED-SL needs to mobilize experts and volunteers who are endowed with these skills. It is extremely difficult for organizations such as GYNED-SL to mobilize external funding to implement climate change adaptation projects because of the large number of international development agencies (bilateral and multilateral and IGOs) operating directly in Sierra Leone. The organization should explore the following options if it is desirous of mobilizing financial resources to support its programme: (a) it should explore the possibility of partnering with one of these organizations working on the ground in Sierra Leone to implement a climate change adaptation project in partnership with these organizations, (b) it should be subcontracted by such organization to implement the project on behalf of these organization, and (c) it should work with the appropriate governmental agencies including the Ministry of Environment and/or the Environmental Protection Agency that is willing to endorse the organization's project so it can access funding from GCF, GEF and other donor agencies. It can also work with the government and other development partner agencies to establish a social investment fund to provide adaptation finance to vulnerable communities (Horstmann & Schulz-Heiss, 2014).

The quality of the GYNED-SL would largely depend on the capacity of the organization's staff to deliver its programmes to the poor and marginalized youth and their communities. Due to the uncertainties of its finances and the fierce competition among development agencies to recruit the small amount of skilled personnel avail locally in the development field, the organization is experiencing difficulties keeping these persons on as staff. Considering the absence of sustained, reliable funding sources to cover the staffing cost, the organization collaborated with universities and colleges across Sierra Leone to recruit volunteers and interns to participate in its programmes. After completing their internship, it will hire these young people to work as staff after completing their academic studies. GYNED-SL implemented its own staff training programmes and "piggybacked" other capacity-building opportunities by partner organizations as strategies for upgrading its staff's knowledge, skills and competencies (Chanita-Mataya et al., 2020).

4.1.9 Methodologies Utilized and Effectiveness in Building the Capacity

The organization utilized various participatory methods, including workshops, mentorship, internship, networking and coaching, to develop the capacity of its staff, volunteers and beneficiaries:

> Workshop and online training are two of the most popular method used to train members of the organisation to resolve the many capacity development challenges confronting the organisation. It also encourages staff and other youths to participate in apprentices and internship programmes provided by credible development organisations such as NAYCOM, MOYA, AVSI and others. GYNED-SL runs an internship programme and usually accepts students desirous of working as volunteers from universities and colleges to participate in programmes. Alma M. E. N. Sheriff, staff of GYNED-SL

Staff testified that they benefited professionally and personally from the capacity development programmes implemented by the organization:

> With regards to work plan, it taught me how to use many methods to plan our work in the future. It was a challenge for me in the beginning, but it is that which gave me the energy to come to work every day and report to my superior. Ostilda C. O. Hughes, operations director, staff member

Sorie I Koroma, staff of GYNED-SL, said, "I am pleased with the new skills I learned at online courses and in weekly capacity development workshops run by the organisation. I used my diary to put down all the activities I will be engaged in the upcoming week (work plan). Now I can see personal improvement in the way I conduct myself and my work and how I implement my personal chores".

GYNED-SL utilized a range of participatory and online methodologies to enhance the capacity of its staff (Chanita-Mataya et al., 2020). On Monday, the organization will implement its weekly staff internal capacity training. These capacity-building trainings were implemented on a needs basis, i.e. if there is a deficiency identified in specific skills required by the staff to implement a programme/project, the management of the organization will invite an expert to deliver the training to boost staff knowledge and skills on the particular development issues. The staff would also identify the capacity-building training and/or internship required through their quarterly training needs assessment exercises, and the organization would recruit an external expert in the field to implement the training (Chanita-Mataya et al., 2020). These training needs assessment exercises are done through one-to-one interviews, focus group discussions and questionnaires (Angelsen et al., 2011; Robson, 2002).

Staff are also encouraged to do online courses in their areas of interest to boost their capacity. The organization supports these staff by paying admission fees for postgraduate and short-term accredited studies. These online accredited courses are very popular with the staff of the organization. The organization also hosts an annual staff retreat to enhance the skills of all staff and volunteers and send them on external training programmes within Sierra Leone and overseas. Staff and volunteers are also sent on internship and volunteering opportunities with bi- and multilateral agencies to boost their capacity in project management, strategic planning and other appropriate development skills. Senior staff in the organization must mentor and/or coach junior staff and volunteers as part of its ongoing mandate to create a learning organization. This strategy facilitated the transfer of skills and knowledge from senior to junior staff (Chanita-Mataya et al., 2020). Junior staff with specific skillset are requested to train and mentor senior staff members within the organization's reverse mentorship development approach. Reverse mentoring is a process that reverses traditional mentoring, usually by pairing junior staff to mentor more senior leaders in an organization (https://www.theconversationspace.com/products/listen-down-2/?src=g).

The staff, volunteers and beneficiaries posit that the combination of bottom-up participatory methodologies implemented by the organization was effective. They said that the methods enhance the capacity of staff, volunteers and key stakeholders to participate in the development programmes implemented by GYNED-SL. The staff and volunteers take advantage of all the opportunities provided by the organization to build their capacities.

4.1.10 Lesson Learned from Implementing the Capacity-Building Programmes

The key lessons learned from implementing the GYNED-SL's capacity-building programmes are as follows:

- The capacity-building programmes and the methodologies implemented by GYNED-SL enable staff, volunteers and beneficiaries to develop unique skills and knowledge on managing their work and time effectively. Staff said they learned how to work smartly and with decency.
- The capacity-building programmes enable the staff of GYNED-SL to communicate more effectively with a wide range of stakeholders involved in the project, including governmental agencies, donor agencies, district representatives, youth and other beneficiaries.
- The project enhances staff learning and knowledge and the commitment and dedication of staff and beneficiaries. The staff are more enthusiastic and dedicated to the development of people and expressed their willingness to participate in the country's development. They said they are more committed to seizing any opportunities to improve themselves and the organization and add value more to the organization's work.
- The online course provided opportunities for staff to enhance their knowledge, skill and competencies in climate change adaptation, social and economic development, strategic planning and other critical areas of development.
- The capacity development training enables staff to develop weekly work plans of activities, assess the achievements of the activities for the past week(s) and plan strategies for the coming weeks based on the evaluation of the overall programme.
- Staff are more skilled in developing and utilizing networks for sharing knowledge and experiences on volunteering, working as a team and implementing projects at district and national levels.

4.1.11 Best Practices Evolving from the Work of GYNED-SL

The best practices evolving from capacity-building programmes implemented by GYNED-SL:

- In terms of project development and implementation, GYNED-SL developed an excellent model for encouraging and facilitating teamwork among staff and key stakeholders. One female staff said, "when there is a project, we do everything together as a team…we check for mistakes in the content of project documents, identify the gaps in the operational processes/procedures before and make the amendments where necessary before submitting the project document to the governments and donor agencies for funding".

- GYNED-SL works are participatory, bottom-up and process-oriented. There are always opportunities for all stakeholders to participate in the organization's governance processes by submitting their own ideas on what outcomes they wish to see when implementing the projects/programmes. The organization instituted mechanisms to ensure that it communicates the outcomes of the projects to all stakeholders.
- The organization's work must always be of the highest quality and standards. Excellence is promoted as a core value of the organization. Staff are encouraged to promote such core values even in their personal life.
- The work of GYNED-SL in youth entrepreneurial development is reaping dividends and possesses the potential to impact research in the field of youth entrepreneurship and sustainable livelihoods. Staff travelled to the field to collect data through questionnaires, interviews and focus group discussions (FGD) on these communities' issues. The data collected in the field is used to shape the programmes developed and implemented by the organization.
- The organization helped staff to enhance efficiency in their work by developing their leadership qualities and trust among staff. One female staff member said she worked with the people she trusts and has confidence in the people she works with in GYNED-SL. Such leadership qualities drive the organization's mentorship, internship and coaching programmes.
- Due to the small staff, the vastness of the work to be done, the vast spread of the communities and the large population of youth to be covered by the GYNED-SL, staff and volunteers are encouraged to practice multi-tasking as an integral part of the organization's work.
- The organization utilizes a participatory development model for engaging beneficiaries and key stakeholders frequently on the overall structures and strategic operations of the organization.
- The organization is committed to working with partners such as TVET centres, colleges and universities to enhance young people's capacity by coordinating and facilitating their participation in training, apprenticeship and internship programmes.

4.1.12 The Recommendations for Improving the Capacity of the Organization

NGOs are the third sector of the economy (Kallman & Clark, 2016). The third sector of society plays a critical role in creating values worldwide through its work in service delivery, advocacy, cultural programmes and social movements. Due to the important roles GYNED-SL plays in the development of Sierra Leone, here are the recommendations to enhance the growth and inclusion of the organization in the development planning processes of Sierra Leone.

- The subject of sustainable finance is a critical challenge for GYNED-SL to achieve sustainability in its programmes and support the country in working towards achieving the UN Sustainable Development Goals (SDGs). Therefore, donor agencies should collaborate with NGOs such as GYNED-SL to finance their projects in the field.
- The policy of the Government of Sierra Leone is to provide incentives and other support to organizations employing more than ten persons; therefore, there is a need for the government to implement such a policy to help GYNED-SL to expand its programmes to more remote communities in Sierra Leone.
- All efforts must be made on the part of the government to maintain and sustain the current government and donor partnership in Sierra Leone. The GYNED-SL will make concerted efforts to approach potential development partners internally and externally to

work with them to develop and implement joint development projects in the field.
- The organization should utilize the media, including traditional and electronic media, to continue to showcase the innovative, dynamic and participatory work it is implementing in marginalized rural and urban communities.
- GYNED-SL should make a concerted effort to recruit more youth volunteers from rural and urban areas and district and community levels across Sierra Leone.
- The organization should explore and promote further innovative capacity-building opportunities to enhance the skills, knowledge and competencies of youth, women, farmers and other key stakeholders who are unemployed, are living in poverty, have poor health and have dilapidated environmental situations in rural and urban communities.

4.1.13 Conclusion for This Chapter on Capacity-Building

Like most LDCs located in West Africa, Sierra Leone has a young population of approximately 60% under 25 (Government of Sierra Leone, 2015). Young people in Sierra Leone encounter varying development challenges, including unemployment, illiteracy, teenage pregnancy, poor health prospects and the trafficking and use of illicit drugs. The limited employable skills, work experience and low educational levels of many young people are responsible for the high levels of unemployment. This situation is compounded by the severe impact of climate change on young and their communities. Sierra Leone also suffers from a severe difference in the human development capacity in its institutions and the inability to resolve the developing challenges facing the country (Simmons, 2021, 2022).

This chapter explored the top-down and bottom-up approaches implemented through the UNDP-funded Government of Sierra Leone "Youth Employment and Empowerment Project (YEEP)" implemented in 2012–2018 and the GYNED-SL youth entrepreneurship development project to build the capacity of institutions and key partners to drive youth development in a post-conflict LDC (Matland, 1995; OECD, 2013). The YEEP focuses on strengthening the capacity of the NAYCOM and the MoYA by building the capacity of both institutions to deliver youth development programmes. Through the YEEP, the "Youth Development Partner Coordination Framework", the development of the blueprint for the youth development programmes 2014–2018, the mapping of donor support to fund youth development projects in Sierra Leone and the provision of youth skill TVET training to enhance the employability of youth. The second case study focuses on implementing the bottom-up approach implemented by GYNED-SL to build the capacity of its staff and young people and their communities in Sierra Leone by utilizing participatory approaches to enhance entrepreneurship and sustainable livelihoods.

Both the top-down and bottom-up approaches have their own strengths and weaknesses. The outcomes of the top-down approaches are summarized as (a) the development of the Youth Coordination Development Framework, which provides a system for coordinating all programmes/projects across the governmental institutions and beyond. The framework was supported by the implementation of the M&E system, which provides feedback to the policymakers on the performance of the programme, (b) offers strategic directions for the government on the specific capacity development training implemented to achieve the developmental policies and (c) advices the government on the particular policies and roadmaps for different development outcomes relating to building the capacity of youth development institutions to enhance the resilience of communities to resolve poverty and adapt to the impact of climate change and other development calamities.

The strengths of the bottom-up approach implemented by GYNED-SL include:

(a) Recognition of GYNED-SL by the government, other development partners and the local communities as the leading youth development organization in Sierra Leone

working on youth employment and entrepreneurial issues.
(b) Staffed by a team of young professionals who are passionate, dedicated and committed professionals and volunteers.
(c) Possesses good working relationship with MoYA and NAYCOM and other governmental agencies working with youth and other marginalized sectors of the population.
(d) Scope of work of the organization is nationally driven by a cadre of volunteers and staff in all districts, utilizing innovative participatory approaches to implement successful youth development projects in collaboration with MoYA, NAYCOM, UNDP, UNFPA, ADB, GIZ and others.
(e) Beneficiaries participate directly in making decisions on the nature of the project they wish to implement in their communities.

GYNED-SL helps young people to become volunteers within their communities by building their capacity through training, internship, mentoring, networking and coaching.

Like any development initiative, both cases indicated their own weaknesses and idiosyncrasies. The informants of the UNDP/Government of Sierra Leone YEEP project state that there was not enough participation of local people in the design, development and implementation. The programmes and strategies implemented were mainly driven by external experts and consultants, while local consultants and experts played a limited role in the project implementation. The implementors of the project did not give much attention to engaging and developing local capacities to implement similar projects in the future. It attributed such challenges to the limited staff composition and competencies in NAYCOM and MoYA in relation to the task to be performed, making it very difficult to build the institution's capacity in a meaningful and strategic way.

The major weakness of the GYNED-SL was its inability to mobilize needed funds to support its programmes, a problem compounded by the magnitude and complexities of the development challenges facing thc youth sector in Sierra Leone. One can deduce from the responses from key informants that the organization has overextended itself by working in all districts in Sierra Leone rather than in a handful of targeted districts that need its intervention. The latter strategy would indicate that rather than stretching itself too thin, the organization would be more strategic in utilizing its limited financial and human resources to resolve the community's development issues.

Like most civil society institutions operating in Sierra Leone, GYNED-SL cannot deliver the portfolio of programmes it is currently implementing. The organization lacks experienced staff and volunteers to effectively deliver its programmes. Due to the financial challenges facing the GYNED-SL, it is experiencing difficulties maintaining and sustaining experienced qualified staff to implement its programmes. It heavily relies on university volunteers and interns to sustain its programmes' implementation.

Both institutions utilized similar capacity-building strategies to enhance staff's skills, knowledge and competencies to deliver sustainable development programmes. The YEEP capacity-building training workshops have as their lead strategy to build the capacity of staff, partner institutions and beneficiaries. The participants who attended these training workshops were provided with reference materials on various topics to boost their competencies. Senior staff of NAYCOM and MoYA were sent on regional and international training courses. These training courses utilized methodologies such as whole-class presentations, panel discussions and group work activities to stimulate and sensitize them and facilitate the transfer of new skills and knowledge, while GYNED-SL utilized a range of participatory and online methodologies to enhance the capacity of its staff, volunteers and youth.

On the issues regarding the effectiveness of the methodologies utilized to build the capacity of staff and participants, the response from key respondents of the UNDP/Government of Sierra Leone YEEP presented differences in the views from very effective to not too effective. Therefore, one can conclude that although learning took place, the methodologies could have been more effective. However, the feedback from the

GYNED-SL staff, volunteers and beneficiaries shows that the blend of bottom-up participatory methods implemented by the organization was very effective and contributed to enhancing the capacity of all.

One key lesson from the Sierra Leone experience is that capacity-building works must be formulated and packaged strategically to be implanted in development institutions. This approach helped the institution sustain its capacities and memories when key personnel left the organization. The capacity development programmes implemented through the UNDP/Government of Sierra Leone top-down YEEP and the GYNED-SL bottom-up capacity-building programmes not only enhance the skills and knowledge of youth to exploit sustainable livelihood opportunities but also contribute to harnessing the leadership and management capacity of other governments and partner institutions.

In terms of the best practice evolved from the capacity-building programme, both initiatives contributed to enhancing the capacity of the staff and beneficiaries of the stakeholders and implementing partners. The staff of these agencies were also involved in all levels of the training and consultations process and acquired new skills and competencies, which contributed to enhancing these institutions' overall management and operation. The enhancement in the capacity of these institutions enables them to better monitor, report, train, track and manage their youth development programme. However, the effectiveness and efficiency of these institutions in sustaining the development and implementation of their programmes have been constantly challenged by the Ebola pandemic, the devastating impacts of climate change, poverty and other developmental calamities which impeded national growth and development. Despite these challenges, beneficiaries utilized the skills acquired to deliver programmes to reduce the impact of the deadly Ebola epidemic, including contact tracers, coordination, communication, awareness building and data management operations. These skills and knowledge gained by beneficiaries contributed to enhancing the capacities of participating agencies and instituted defensive remedial measures, which reduced the impact of the COVID-19 pandemic on the people of Sierra Leone.

The participatory nature of the bottom-up approach implemented by GYNED-SL warranted a special mention. The organization provided opportunities for beneficiaries to participate in the governance and management processes of the organization by creating fora for them to put forward ideas on what outcomes they wish to experience from projects implemented by the organization. The organization instituted mechanisms to ensure that it communicates with all stakeholders involved in the project development process. It adapted and promoted excellence as a core value of the organization.

4.2 Youth Empowerment and Employment Programmes (YEEP) in Sierra Leone, 2012–2015: Top-Down Approach to the Implementation of the UNDP/Government of Sierra Leone

4.2.1 The Socio-economic Situation of Sierra Leone

Unemployment has increased rapidly over the past 3 years due to global economic challenges and the COVID-19 pandemic. Globally, approximately 679 million people are living in extreme poverty. About 9% of the world's population works for less than $1.90 per day (https://devinit.org/924633#c6ee5bea), and according to UNICEF, around 22,000 children die from hunger each day (https://globalissues.org/article26/) despite the wealth generated. In Sierra Leone, young people account for approximately 62.5% of the national population (under 25 years), estimated at 7.97 million (https://sierraleone.unfpa.org/en/node/6136). Youth in Sierra Leone face numerous development challenges, including securing decent employment in an economy characterized by a lack of job opportunities, especially in rural communities. The limited employable skills, work experience and low educational levels of many young people are responsible for the unemployment situation they are experiencing. The country is also trying to recover from the effects of an 11-year civil war that ended in 2002 and, more recently, the devastating Ebola pandemic.

The youth unemployment situation in Sierra Leone is characterized by poor working conditions compounded with long hours and low service pay. The youth also encountered mounting difficulties of poor-quality education, poor health and nutritional standards and increases in HIV/AIDS, teenage pregnancy, drug abuse and violent crimes. Approximately 70% of the youth population of Sierra Leone is either underemployed or unemployed, and 50% are classified as either illiterate or unskilled (UNDP, 2015; ILO, 2010, 2013). Although there are vast improvements in the situation of youth during the period under review, when compared to the situation before the civil war, many pre-conflict conditions that helped fuel the youth's participation in the civil war still exist today. Large numbers of unemployed youths are a potential source of insecurity, given their vulnerability to recruitment into criminal and violent activities. Young women faced high unemployment in both rural and urban areas. These were the predominant development conditions responsible for the UNDP and the Government of Sierra Leone to develop and implement the US$ 9 million Youth Employment and Empowerment Programme (YEEP).

4.2.2 Youth Employment and Empowerment Programme (YEEP)

This project is implemented through the following strategies: (a) provision of employment opportunities and income-generating activities for young people, (b) strengthening the capacity of the Ministry of Youth Affairs (MoYA) and National Youth Commission (NAYCOM) to deliver sustainable livelihoods projects, (c) promotion of the participation of young people in decision-making processes and (d) the development and implementation of youth interest development initiatives.

4.2.3 The Aim and Objectives of the YEEP

The programme aims to enhance institutional capacity development, policy development and youth employment through business development services and support for "school-to-work" transition and employability, including Career Advising and Placement Services and Graduate Internship Programme.

The objectives of the programme are:

1. Provide rapid employment opportunities and income-generating activities for young people.
2. Strengthen the capacity of the Ministry of Youth Affairs (MoYA) and National Youth Commission (NAYCOM).
3. Promote and facilitate the participation of young people in decision-making processes and the development of youth interest initiatives.

An important element of the YEEP initiative is strengthening the youth development coordination structure and leadership of MoYA and NAYCOM. This action aims to improve national youth development institutions' coordination, strategic leadership and oversight. The Youth Employment and Empowerment Programme is part of the "Inclusive Growth and Sustainable Development Programme Cluster of the Sierra Leone Country Programme Document and Country Programme Action Plan" (https://www.sl.undp.org/content/sierraleone/en/home/about-us.html#:~:text=UNDP%20supports%20the%20people%20of,and%20democratic%20insitutions%20and%20processess).

The UNDP has the experience and technical competencies to develop employment generation and private sector development initiatives in Sierra Leone in the post-conflict period. It has championed the creation of income-generating opportunities for young people across a broad array of activities ranging from micro-enterprise, financial literacy and inclusion training and business development schemes to support the development of cooperatives and promote access to finance.

UNDP has supported the establishment of NAYCOM. Through the NAYCOM framework, UNDP has provided young entrepreneurs with a broad range of business development support services by establishing five business support centres (BSCs) in Freetown, Newton, Bo, Makeni and Kenema. These BSCs provide direct assistance to over 2000 young entrepreneurs per annum.

4.2.4 Coherence of YEEP Within the UN Joint Vision and PRSP II

The UNDP is a United Nations institution that works in more than 170 countries and nations to eradicate poverty and reduce inequalities through the sustainable development of nations (https://www.undp.org/). YEEP initiative was aligned with the Second and Third Poverty Reduction Strategy (PRSP II and 111): 2008–2012. It also forms part of the UN Joint Vision Programme 19, which has a single framework for the coordination of all youth interventions supported by the UN. The UN Joint Vision Programme 19 forms part of the "Joint Response to Youth Employment in Sierra Leone". The programme includes the creation of fora for policy and strategy dialogue between the government and all development partners working on youth development issues (Government of the Republic of Sierra Leone, 2010).

The UNDP's work in Sierra Leone centres on two strategic areas: (a) Inclusive Growth and Sustainability and (b) Effective and Inclusive Democratic Governance. In Inclusive Growth and Sustainability, the UNDP SL works with the people of Sierra Leone to reduce poverty and marginalization – regarding the most vulnerable and excluded economically, socially and environmentally sustainable ways. The Effective and Inclusive Democratic Governance strategic programme supports the people in creating an enabling environment for stability, the rule of law and good governance. The strategy includes strengthened service delivery, legal and judicial systems and democratic institutions and processes.

(https://www.sl.undp.org/content/sierraleone/en/home/about-us.html#:~:text=UNDP%20supports%20the%20people%20of,and%20democratic%20insitutions%20and%20processess).

A critical component of the YEEP initiative is strengthening the youth development coordination system of MoYA and NAYCOM. This strategy aimed at bringing better coordination, oversight and strategic leadership to the national youth development institutions. The Youth Employment and Empowerment Programme is part of the "Inclusive Growth and Sustainable Development Programme Cluster of the Sierra Leone Country Programme Document and Country Programme Action Plan" (CPD and CPAP).

4.2.5 The YEEP Project

The YEEP initiative is aligned with the PRSP II "Agenda for Change" programme. It also forms part of the UN Joint Vision Programme 19, which has a single framework for the coordination of all youth interventions supported by the UN. The UN Joint Vision Programme 19 forms part of the "Joint Response to Youth Employment in Sierra Leone". The programme comprises fora for policy and strategy dialogue between the government and all development partners working on youth development issues.

The YEEP project is implemented in partnership with the Ministry of Youth Affairs (MoYA) and the National Youth Commission (NAYCOM). The project's overall aim is to support MoYA and NAYCOM to empower the youth to develop their potential, creativity and skills for national development. Besides the enhancement of the institutional capacity of MoYA and NAYCOM, the components of the YEEP include (a) Business Development Services (BDS), (b) Career Advising and Placement Services (CAPS), (c) Graduate Internship Programme (GIPS) and (d)

Agriculture Skills Development Programme (Simmons, 2022).

During the period under review, YEEP has become the foremost youth capacity development programme implemented in Sierra Leone geared specifically at empowering the youth sector to participate effectively in the national development processes and programmes. The programme strengthened national policy, strategy and coordination frameworks for youth employment. It was also geared to establish essential support services for youth, such as (a) Business Development Support (BDS) and (b) Youth Career Advice Services (YCAS), which are geared towards preparing youth for the employment market.

YEEP has capitalized on the following strategies to fast-track the preparation of youth for the employment market and to empower them to participate in the decision-making process, namely, (a) provision of rapid employment opportunities and income-generating activities for young people, (b) strengthening the capacity of the MoYA and NYCOM, (c) promotion of the participation of young people in decision-making processes and (d) the development of youth interest initiatives.

The YEEP project is part of the "Inclusive Growth and Sustainable Development Programme Cluster of the Sierra Leone Country Programme Document and Country Programme Action Plan" (CPD and CPAP). To achieve inclusive growth and sustainable development now and into the future, as stipulated in the Country Programme of Action, the YEEP initiative is geared towards strengthening the youth development coordination system, policy and implementation frameworks of MoYA and NAYCOM to bring better coordination, oversight and strategic leadership to the national youth development institutions.

4.2.6 Overview of YEEP Sub-programmes

The components of the YEEP are (a) the Business Development Service (BDS), (b) the Career Advising and Placement Services (CAPS) centres, (c) the Graduate Internship Programme (GIP), (d) Sierra Leone Agricultural Business Initiative (SABI) and (e) the Policy and Institutional Development Programme.

4.2.6.1 The Business Development Service (BDS)

The Business Development Services (BDS) initiative is the component of the YEEP geared towards preparing young people for self-employment through developing enterprises and the employment market. Its aim is to create and expand sustainable enterprises leading to increased employment opportunities for young people in major urban and rural centres. In 2012, five business support centres (BSCs) were established in the main urban centres in Freetown, Newton, Bo, Makeni and Kenema. These centres provide training, direct assistance and support to over 2000 entrepreneurs annually (Simmons, 2022).

4.2.6.2 The Graduate Internship Programme (GIP)

GIP was established in 2012 as a business model for successful partnership and cooperation among various development partners. UNDP and NAYCOM created a successful partnership with over 68 governmental institutions, civil society organizations and private businesses in Sierra Leone. The GIP provided training and internship opportunities for young people to build critical work ethics and practices through on-the-job experience/internships. For the duration of the YEEP implementation process, 2323 youth have applied for the GIP initiative. Of that amount, 1137 young people have successfully participated in the programme. Although the programme has been a tremendous success in terms of the number of persons who successfully completed the initiative, the major shortcoming of the GIP was attracting young women. Only 28.5% of the young people who successfully completed the initiative were young women (324).

4.2.6.3 The Career Advising and Placement Service (CAPS) Centres

CAPS were designed to equip university and college students with (a) labour market and career development information, (b) job search skills, (c) information technology (IT) training, (d) job placement opportunities and (e) career development workshop opportunities. CAPS aimed to foster constructive cooperation and relationships between tertiary-level educational institutions and potential employers. Five CAPS centres were established. These centres were attached to major academic centres, namely, Fourah Bay College (FBC) in Freetown and Njala University in Bo, offering employment support service skills and life skills to improve the employability of young people as they complete their academic training (Simmons, 2022).

4.2.6.4 Sierra Leone Agricultural Business Initiative (SABI)

The SABI initiative is an agri-entrepreneurship scheme modelled from the association with the Songhai Centre in Benin in 2012. The Sierra Leone Agricultural Business Initiative (SABI) is aimed at boosting food production by (a) making agriculture and agribusiness attractive to youth entrepreneurs and (b) creating employment opportunities for rural and urban youth to earn sustainable livelihoods. The SABI strategy was achieved through (a) the implementation of integrated training for youth, (b) the transformation of the agricultural value chain into more inclusive and value-added chains, (c) the introduction of technological innovations into agriculture value train to create sustainable production, (d) the introduction of and development of mechanization, (e) the marketing promotion of new and existing products and services and (f) the introduction and promotion of recycling services.

4.2.7 The Policy and Institutional Development Programme

A National Youth Programme, 2014–2018, was launched and implemented in 2014. A draft National Youth Policy (NYP) was developed, and its consultation process was completed. The National Youth Programme, 2014–2018, mandates MoYA/NAYCOM to collaborate with various governmental agencies, INGOs and development partners, civil society agencies and private sector agencies involved in youth development to harmonize the coordination of youth development programmes across the sectors. The NYP mandated MoYA/NAYCOM to learn from best practices of what has worked in other sectors in Sierra Leone: the Justice Sector Coordination Office, the Health Sector Joint Implementation Unit and the Scaling-Up Nutrition Cross-Sector Coordination Secretariat. It also recommended that MoYA/NAYCOM implement four coordination mechanisms to enable them to successfully coordinate these programmes (Fig. 4.2).

The coordination mechanisms of the Youth Development Partner Coordination Framework comprise:

Coordination Mechanism 1: Inter-ministerial Committee on Youth Empowerment and Development

Mechanism 1 consists of government ministers and permanent secretariat of the governmental ministries who are responsible for the development and implementation of governmental policies and programmes which support youth development programmes. Heads of international development organizations (IGOs), including UNDP, GIZ and others, working on youth development issues would participate in these high-level meetings. The mechanism is chaired by the president and/or his designate twice per year.

Coordination Mechanism 2: National Youth Policy Steering Committee (NYPSC)

Mechanism 2 is chaired by the minister of youth and/or his/her designate and would ensure the development and implementation of a relevant national youth policy and supported strategic programmes within the Ministry of Youth and the National Youth Commission. The mechanism is comprised of senior officials from the Ministry of Youth Affairs (MoYA) and the National Youth Commission (MoYA) working in collaboration with NGOs and IGOs on the development and

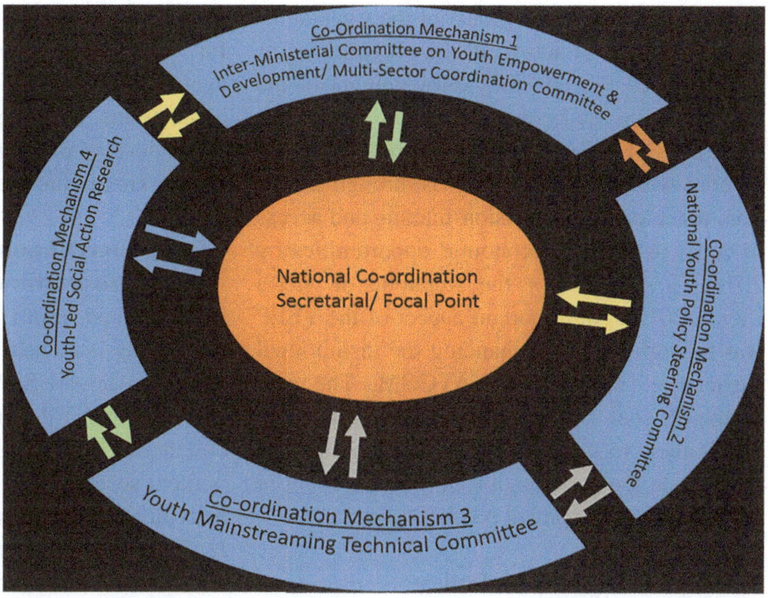

Fig. 4.2 Coordination mechanisms of the Youth Development Partner Coordination Framework

implementation of youth development programmes in Sierra Leone.

Coordination Mechanism 3: Youth Mainstreaming Technical Team

The mechanism comprises senior technical officers from key governmental ministries and divisions, the representative of the National Youth Council (NYC) and IGOs whose programmes are critical to the implementation of the National Youth Policy and the youth mainstreaming initiatives across governmental ministries, divisions and partner institutions. The main function of this mechanism is to ensure the mainstreaming of youth development in terms of programmes and sorted by effective allocations of resources across key ministries and sectors. The representatives from the various development institutions and governmental ministries and divisions would report to the mechanism on the performances of their respective youth mainstreaming programmes.

Coordination Mechanism 4: Youth-Led Social Actions Research Network

Mechanism 4 is chaired by the deputy youth minister and/or youth commissioner and consists of senior technical and programme officers and researchers from the appropriate governmental department, tertiary institutions, NGOs and IGOs. This mechanism is responsible for the development of the National Youth Policy and National Youth Programmes and conducts extensive research on various aspects of the national youth development programmes to ensure their relevance and effectiveness.

The implementation of the coordination mechanism was disrupted by the outbreak of the devastating Ebola epidemic in 2014–2015. This was also compounded by severe tensions between the leadership of MoYA and NAYCOM, which heavily impacted efforts to reactivate the implementation of the coordination mechanism. General elections were called in 2018, and the government was replaced by a new political party with a new development agenda for the country. The UNDP YEEP ended in 2018.

4.2.8 Youth Policy and Youth Development Coordination Mechanism in Sierra Leone

4.2.8.1 Agenda for Prosperity

In Sierra Leone, the "Agenda for Prosperity" (2013–2018) succeeded the Agenda for Change and placed youth employment as a core outcome in the new development agenda. For the youth development initiative to be successful, the youth development agenda of Sierra Leone must be integrated within all the pillars of the Agenda for Prosperity and cut across all the programmes of all ministries and divisions (Government of the Republic of Sierra Leone, 2013).

Pillar 8 of the Agenda for Prosperity focuses on empowering women and girls through education, participation in decision-making and access to equal justice and economic opportunities by 2018 (Government of the Republic of Sierra Leone, 2013). An important aspect of the YEEP was to work towards enhancing the institutional capacity of MoYA and NAYCOM. The programme's goal is to strengthen these agencies to enable them to successfully deliver programmes to enhance the participation of young people in the national development process.

4.2.8.2 National Youth Policy

The vision of the National Youth Policy (2010) was to develop young people who are nationally conscious and empowered to contribute positively to the development of Sierra Leone (Government of the Republic of Sierra Leone, 2010). The National Youth Policy aims to create an enabling environment where youth development and empowerment programmes can sustainably realize desired long-term goals and outcomes. The policy is based on the Commonwealth Plan of Action for Youth Empowerment Principles, 2007–2015. It utilizes a youth rights approach to youth development, which seeks to mainstream the rights of youths as exemplified in the Constitution of Sierra Leone, the National Education Policy, the UN Rights of the Child and the Lisbon Protocol of 1998 (Commonwealth Secretariat, 2007; Government of Sierra Leone, 2020).

The National Youth Policy is focused on addressing major challenges facing young people in Sierra Leone, including youth unemployment and underemployment, lack of access to quality education and inadequate preparation and/or training for the formal job market, drug and substance abuse, failure to address gender issues among young people and sexual and reproductive health issues.

The policy also provides a framework for collective action, implementation and coordination, monitoring and evaluation (M&E) arrangements for youth development among governmental agencies, youth-serving organizations, youth associations, civil society organizations, international development partners and other stakeholders working in the youth development sector.

4.2.8.3 Sierra Leone National Youth Programme, 2014–2018

The National Youth Programme (NYP), 2014–2018, was launched in April 2014 and is estimated to cost US$217 million over the next 4 years. The NYP mandates MoYA/NAYCOM to collaborate with key agencies in the youth development sector to streamline the coordination of activities across the sector. It instructed MoYA/NAYCOM to learn from best practices in other sectors and transfer them to inform the programming and coordination in the youth development sector.

4.2.9 Methodologies

Multi-data collection strategies were utilized, which included desk audit and review of documents, questionnaires, interviews and participatory rapid/rural appraisal techniques, including focus group discussions, to collect data for the case study (Table 4.1).

4.2.10 Semi-structured Interviews

Semi-structured interviews were used with senior managers from NAYCOM, UNDP and MoYA and other relevant stakeholder ministries and partner agencies to provide a fair, reasonable and credible assessment of how well YEEP was understood and whether it responded relevantly to the development context of Sierra Leone (Saunders et al., 2000; Robson, 2002).

The semi-structured interviews were focused along the lines of the OECD-DAC evaluation criteria (OECD/DAC, 2019) on (a) the assessment of YEEP's performance using the following set of criteria, effectiveness, efficiency, relevance, sustainability, the possibility of duplication, part-

Table 4.1 Matrix indicating the summary of research methods implemented and the benefits and limitations of these methods

Types of method	Application of method	Benefits of method	Limitations of method
Participatory method	Participatory appraisal is an approach that seeks to build community knowledge and encourages grassroots action. It uses many visual methods, making it especially useful for participants who find other methods of participation complicated A facilitated process involving a group of beneficiaries in which members of the group interact, mainly around visual ways of expressing their opinions and thoughts	Useful in answering questions of how and why A mutual learning environment can help build stakeholders' capacity Able to capture a diversity of perceptions Enhanced ability to understand complex processes Good for a general impression of progress or outcomes Ability to capture negative or unintended consequences Can help to identify and articulate people's felt needs Enhances the organization's accountability to its beneficiaries	Costly and time-consuming Requires specialized facilitation and knowledge of appropriate methods to engage people A commonly encountered problem is that as participatory appraisal uses very accessible tools, it is often used as an information-providing exercise that does not follow through to facilitate decision-making within the community
Interviews	Interviews are generally structured so that the interviewees will give their answers to specific questions Interviews can also be less structured if the issues in question do not have recognized answers	Flexible/adaptable Can give visual cues Can combine open questions with pre-coded ones	Possibility of manipulation by the interviewer Vulnerable to personality conflicts Required skilled interviewers Might be difficult to summarize the findings Interviewers may have to travel to collect data; therefore, cost or safety issues are critical

(continued)

Table 4.1 (continued)

Types of method	Application of method	Benefits of method	Limitations of method
Focus groups	Collect data through group interaction on a topic determined by the author/researcher. Often helps to generate questions but not necessarily definitive answers Findings need to be compared to a larger survey. The facilitator's skills can strongly affect the value of a focus group	Group interaction Group consensus In-depth discussions Can be more efficient than one-to-one interviews Use fewer resources than one-to-one feedback Democratic process where the researcher is outnumbered by participants Provide a relatively immediate sense of results	Small sample size Group may not be representative Responses all depend on one another, and group format may create conformity where differences are suppressed May cause people to feel like they need to take sides (polarization) People may be manipulated by others in the group Questions may not be asked the same way each time Difficult to quantify the results or findings Not appropriate for some sensitive issues
Video, film	Indicates the achievements of a project compellingly	Performance-based activities Vivid impression Creative and artistic Can be motivating or fun for participants	Time-consuming Impression of the project or its outcomes can be affected by the quality of filming and presentation rather than the quality of the project Depends on the skills of the viewer in interpreting Taken alone, inability to enquire of participants

nership, community involvement and political commitment, (b) providing a credible assessment of how well YEEP was understood and whether it responded relevantly to the context and (c) identifying the main achievements both expected and unexpected of YEEP, particularly as they relate to the employment promotion and institutional capacity-building, and what facilitated the achievements (https://www.oecd.org/dac/evaluation/daccriteriaforevaluatingdevelopmentassistance.htm).

Due to the nature of the YEEP case study, we also employed semi-structured Interviews to solicit the views of senior-level staff of key stakeholder agencies involved in the delivery of the YEEP sub-programmes, including the Business Development Services (BDS), Career Advising and Placement Services (CAPS), Graduate Internship Programme (GIP) and Skills for Development Programmes (SDP), particularly as they relate to employment promotion and institutional capacity-building aspects of the project and also to identify the factors which facilitated their achievements (Saunders et al., 2000; Robson, 2002).

4.2.11 Focus Group Discussions

The consultants used focus group discussions (FGD) to assess beneficiaries' views to identify the main achievements, both expected and unexpected, of the components of the YEEP project, including BDS, CAPS, GIPS and SDP, particularly as they relate to employment promotion and institutional capacity-building and to identify the factors which facilitated the achievements of the project. A set of questions was identified along the OECD-DAC evaluation criteria listed above (Saunders et al., 2000; Chambers, 2002; Robson, 2002).

4.2.12 Production of Film on the YEEP Initiative

The authors reviewed a short documentary film of 12 min duration on the components of the YEEP project, including BDS, GIP, SDP and CAPS. The film presents an overview of the YEEP project, stresses the development challenges facing Sierra Leone and highlights the consequences of these challenges on young people. It (a) highlights the strategies employed by the YEEP project to engage young people in the national development processes, (b) examines how the YEEP is duplicated and replicated across other stakeholder agencies, (c) focuses on the sub-programmes implemented by YEEP in terms of the mode of delivery and operation and impacts on the lives of young people and their communities and (d) captures the voices of key informants and officials including the youth commissioner, the UNDP official with responsibility for the delivery of the project and the young people themselves, the beneficiaries of the programmes. The film also highlights the best practices of the YEEP in institutional capacity strengthening and management and the factors responsible for their achievement. The film was used to promote YEEP in Sierra Leone and internationally and for training and demonstration purposes.

4.2.13 Validation Workshop

A successful 1-day validation workshop was held in Bo on October 7, 2015. All stakeholders were represented at the workshop. The purpose of the validation workshop was to present the ongoing process of material generated from the data collection process and to obtain feedback and opinions from the strategic partners and key stakeholders to generate the final version of the report. Participants were provided with the draft YEEP evaluation/best practice report. At the workshop's opening, there were presentations from MoYA, NAYCOM and UNDP representatives. The consultants made two presentations to the workshop: (a) the findings evolving from the OECD-DAC approaches and (b) the achievements, lessons learned and recommendations of YEEP. Participants then gather into working groups to discuss the findings of the YEEP report. The draft evaluation/best practice report was endorsed by all participants (https://www.oecd.org/dac/evaluation/daccriteriaforevaluatingdevelopmentassistance.htm).

4.2.13.1 Stakeholders and Their Perspectives

The quality of the participation of the stakeholders in the YEEP is critical to the successful performance of the capacity development project. It is also critical to understand the nature and the role of the YEEP stakeholders to effectively develop and implement viable strategic programmes and approaches to meet the specific needs of the stakeholders involved in the project. There are over 100 stakeholder groups involved in developing and delivering the YEEP project. Below is an analysis of the different stakeholder groups involved in the project, their interests and motivations and what they brought to the programme.

The stakeholder's analysis in Table 4.2 provides valuable information on the stakeholder's interest and how they are affected by the skills deficiency problems in Sierra Leone, their capacity and motivation to bring about changes and the actions to be implemented to meet the interest of the stakeholder. Through the information provided by the matrix, the authors designed the capacity development strategic programmes and approach to enhance the capacity of the staff of MoYA and the NYCOM, the youth training and internship programmes and strategies for strengthening the governance and management of partner agencies.

4.2.13.2 The Strengths of the Organization/Programmes

Due to the human development challenges facing countries in Africa, Dr. Molla McKennon, project manager, UNDP, posits that:

> One of the main strengths of the YEEP was the effectiveness of the strategies used to address the human development challenges of unemployment

Table 4.2 YEEP stakeholders' analysis

Stakeholder and basic characteristics	Interest and how they are affected by the problems	Capacity and motivation to bring about change	Possible actions to address stakeholders' interest
(a) *Community* Household/young people (participants) Project implemented in Freetown, Bo, Makeni and Kenema The communities in Sierra Leone are faced with high levels of unemployment, underemployment, HIV/AIDS and other health challenges and a high incidence of poverty among youth and the wider population	Improvement – Wishes for improvement in their livelihoods HIV/AIDS and Ebola are on the increase, compounded by high incidents of unemployment, crime, illiteracy, poverty, violent crimes, teenage pregnancy and violence among youth	Keen interest in exploring innovative strategies to increase their livelihoods Explore mechanisms to enhance and sustain their livelihoods Lack of the necessary funding and support to enhance their situation	Support to enhance the capacity of youth and CBOs to organize and lobby the government, funding agencies, etc. Identify/develop alternative income sources for women, men and young people such as small businesses, agribusiness, etc. Willingness to participate in business development, technological ventures and other innovations to enhance their possibilities and capacities for employment and sustainable livelihoods

(continued)

Table 4.2 (continued)

Stakeholder and basic characteristics	Interest and how they are affected by the problems	Capacity and motivation to bring about change	Possible actions to address stakeholders' interest
(b) *Government* MoYA was created in 2013 with responsibility for the youth development and empowerment portfolio It is primarily responsible for developing, coordinating, monitoring, evaluating and supporting implementing policies and strategies that promote youth development It lacks the institutional capacity and financial resources to effectively do its job *NAYCOM* NAYCOM was established in 2009 to implement policies, programmes and projects which empower youth to develop their potential for national development (National Youth Programme 2014–2018)	High interest in ensuring that the challenges facing young people are resolved and that they are motivated and empowered to participate actively in national development Lack of financial and human resources capacity High interest in ensuring that young people's challenges are resolved; they are motivated and empowered to participate actively in national development processes Possesses the capacity to implement such programmes Develop a good track record of attaching donor/aid funds because of enhanced capacity in management, implementation, monitoring and evaluation	Headed by a highly motivated and committed minister High political influence in effecting policies and regulations to address youth issues but is not utilizing its influences enough to bring about change MoYA has more interest in accessing donor funds to implement projects Keen interest in collaborating with UNDP and other donors/partners to enhance its capacity by developing management and coordination mechanisms and appropriate policies and programmes to enable it to carry out its functions Highly motivated team of professionals headed by a commissioner with excellent leadership capacity Possesses the capacity to influence/change policy and regulations pertaining to youth development Possesses competencies and capacity to implement youth development programmes/ projects across the country and at all levels	Mobilization of resources and partners from donors, INGOs, local NGOs and institutions to implement innovative youth development programmes. This action will put MoYA in direct conflict with NAYCOM Continue to lobby the central government for more resources (financial and human capacity) to effect policy changes and mainstreaming Implement youth mainstreaming strategic programmes at all levels within government and across all sectors Advocate for and lobby governmental institutions to allocate resources to support youth development actions Strengthen its collaboration with MoYA to enhance the capacity of the ministry to manage, monitor and evaluate structures relating to policy development and youth mainstreaming Enhance capacity to mobilize national and donor resources Collaborate with MoYA to raise the awareness of policymakers, senior officials and the public on the critical nature of youth development in national development

(continued)

Table 4.2 (continued)

Stakeholder and basic characteristics	Interest and how they are affected by the problems	Capacity and motivation to bring about change	Possible actions to address stakeholders' interest
(c) *Central government* Offices of the President and Min. of Finance, etc. Responsible for examining, coordinating, advising and managing all government public expenditure and revenue proposals. Approval of the UNDP and other donor aid projects, including YEEP	High political interest in reducing the incidence of youth unemployment, disempowerment, poverty, HIV/AIDS, teenage pregnancy and violent crimes among the youth Youth development challenges have severe implications for retarding national growth and development	High political interest but lack the political will, resources and motivation to address the country's youth development challenges	Continue to influence the political directorates and the powers that be on the critical importance of engaging youth in national development Continue to mobilize donor aid funds and the allocation of local financial resources to finance the expansion of youth innovative development programmes such as BDS, CAPS and GIPEffect appropriate policies and programmes to enhance the status of youth in Sierra Leone Provide incentives to the private sector, such as tax breaks, etc., to encourage them to employ young graduates of the BDS, CAPS and GIP programmes
(d) *UNDP* In 2011, UNDP implemented a comprehensive development programme, YEEP, aimed at strengthening the national policy, the strategy and the coordination for youth employment and empowerment by seeking to provide basic support services for youth across the country	High interest and commitment to assist the government in addressing the youth employment and empowerment challenges prioritized by the Government of Sierra Leone (GoSL), which is central to "Agenda for Prosperity (2013–2018)"	Highly motivated and possesses the resources, competencies and capacity to enhance the institutional capacities of MoYA and NAYCOM to deliver effective and efficient quality youth development programmes/projects to resolve the employment and empowerment challenges facing Sierra Leone	Sharing of experiences and lessons learned Provide resources and technical support to NAYCOM to up-scale, expand and roll out the BBS, CAPS, GIP and other youth development initiatives Continue to provide resources and technical assistance to MoYA to enhance its capacity to effectively implement its policy development, monitoring and coordination and youth mainstreaming roles Working in partnership with the government of Sierra Leone to organize actions to encourage donors and the private sector to fund projects to support the youth sector

(continued)

Table 4.2 (continued)

Stakeholder and basic characteristics	Interest and how they are affected by the problems	Capacity and motivation to bring about change	Possible actions to address stakeholders' interest
(e) *Other donors and INGOs* Many donor/aid agencies – bilateral and multilateral – work in Sierra Leone to assist the government with its peace and recovery programmes Including (a) UN agencies such as UNICEF, UNIFEM, WHO and UNAIDS and other multilateral agencies such as the World Bank and the EU and (b) bilateral agencies, including USAID, DfID, GIZ, CIDA, SIDA and others working on supporting long-term and equitable economic growth and advancing their nations' foreign policy objectives by supporting economic growth, agriculture, trade, global health, democracy, conflict prevention and humanitarian assistance	High interest in ensuring that project outcomes are achieved in relation to the funding provided More effective management of the resources and outcomes in the youth development sector	Have financial and technical resources to support ongoing research and programme Limited political influence to bring about change on the ground (some have more influence than others)	Provide resources to implement outcomes in the youth development sector and the wider population Sharing of best practices and lessons learned with other countries
(f) *NGOs and CBOs* Numerous NGOs and CBOs deliver education, community development basic needs and relief programmes across Sierra Leone at national, district and chieftain levels	Highly motivated and have a good grasp of the issues and challenges facing the area. Members of agencies are also directly affected by the issues and challenges their organizations hope to resolve	Lack the financial resources to tackle the problem but have the local knowledge and are trusted by the people/communities affected Possess some capacity and local knowledge to solve the problem	NAYCOM and MoYA should collaborate with these agencies to resolve the community's youth development problem by providing resources and technical support to enhance the capacity of the agencies to deliver the programme Collaborate with agencies regarding the development of local promotions and campaigns The possibility of duplication of efforts and wastage of resources if there is an absence of collaboration between government, donors and INGOs

(continued)

Table 4.2 (continued)

Stakeholder and basic characteristics	Interest and how they are affected by the problems	Capacity and motivation to bring about change	Possible actions to address stakeholders' interest
(g) *Training and education institutions and financial agencies* These stakeholders consist of universities, colleges and other technical vocational education programmes offering training in areas of employment and livelihoods, including: The provision of a range of training opportunities in developing the entrepreneurship skills of youths Provision of mentorship and coaching Access to finance	High interest in utilizing their facilities and human resources to provide additional training to graduates to prepare them for employment/job market Frustration from graduates who are still unemployed after completing training from these institutions	Opportunity for institutions to pilot new and innovative areas of training methodologies to prepare graduates for the job market Training institutions have the skills, competencies, human resources capacity and facilities to deliver training, internship and outreach support Additional financial resources in their coffers to purchase new tools and equipment/use of new technology	Delivery of training, internship and outreach support to graduates of the BDS, CAPS and GIPs Enhancing their curriculum to make them more responsive to the job market – Mobilize additional resources to expand and sustain the programmes where possible

> and poverty in Sierra Leone. The UNDP-led YEEP utilised the best practices and lessons learned from similar successful interventions in Africa to resolve the youth employment challenges facing Sierra Leone.

A critical outcome of the programme is its ability to enhance the local capacities in Sierra Leone to build the resilience of the institutions working in the youth sector, mainly working with young people to develop decent employment:

> The outcomes of the YEEP resulted in the enhancement of knowledge and skills within the MOYA and NAYCOM to develop and implement capacity development programs in Sierra Leone, concludes Mumudo Wudie, programme manager of NAYCOM.

The project was timely because it came when the country's civil service and democratic system were destroyed by a civil war that lasted a decade. Young people were perpetrators and victims of the violence inflicted on the people by the civil war. The Government of Sierra Leone was trying to rebuild its democracy and governance system by working with the UNDP and other international development partners to implement the project, which focuses on building the knowledge and skills of youth for employment.

The YEEP contributed to enhancing the officials' skills of MoYA and NAYCOM in programme development and management, monitoring, evaluation and reporting, conflict mediation and resolution. It also enhances the project/programme planning, analytical, leadership and team management skills of the staff of both institutions and other partner institutions participating in the project.

UNDP provided a dedicated fund that finances the capacity-building component of the YEEP over a sustained period. The finances enable UNDP and the government to give a team of consultants and technical experts on the ground in Sierra Leone to support the implementation of the various components of the programme. The range of capacity-building strategies implemented by the YEEP was diverse, effective and appropriate to the country's development needs, including training workshops, university long- and short-term courses, networking, internship, coaching and mentorship.

4.2.13.3 Weaknesses

Although the project was deemed successful from the reviews and evaluations conducted by external evaluators on behalf of UNDP and the

Government of Sierra Leone, the project has its own weaknesses.

> It does not build on area-specific or locally practised indigenous technical knowledge of the targeted communities. This could have helped to build on existing core capabilities of the local communities at district and chiefdom levels, stated Dr. Molla McKennon, former YEEP project manager, UNDP.

He concluded that the interventions implemented by UNDP were mainly driven by experts' interests without giving much attention to local capacities. The NAYCOM has a very limited staff capacity in relation to the task it must perform, making it very difficult to accomplish the mandate of the institution. This situation was compounded by the magnitude and complexities of the development challenges facing the youth in Sierra Leone.

Numerous logistical issues confronted the efficient operationalization of the youth development programme. These issues include the unavailability and the inefficient use of transportation, the low capacity of staff to implement the youth development programme, limited materials and financial resources to support the implementation of the programmes, and the remoteness of some communities from the capital, Freetown, adds to the complexity of logistical challenges encountered.

4.2.13.4 Challenges

Sierra Leone struggled with the impact of a decade-long civil war that destroyed most of the physical infrastructure, governance and democratic systems. The development challenges included poor governance systems and structures, low levels of awareness of the beneficiaries and government officials of the development challenges facing the country and the strategies to be implemented to resolve these challenges.

The exclusion of the youth from participating in developmental opportunities was viewed by informants as a critical issue which regards the MoYA and NAYCOM from accomplishing their development objectives. Although efforts were made by the government to develop the institutional capacity of NAYCOM and MoYA, the low capacity of the staff who are saddled with the development of these institutions makes it extremely difficult to streamline the institutions' programmes with the anticipated development goals of the government.

The lack of and/or low level of political commitment of the Government of Sierra Leone to tackle the development challenges facing youth in the country was viewed by informants as a significant development challenge. This problem was demonstrated in two ways: (a) ethnic-based biases resulting in only youth from the capital and the northern districts being provided with resources and opportunities while those from the southern districts were marginalized and/or isolated and (b) the non-participatory nature of policy and programme development where youth and marginalized sectors of the society were not provided with the opportunity to participate in the development of youth programmes. These problems are compounded by the low levels of human and financial resources available to the government to implement significant programmes at the national, district and chiefdom levels.

The UNDP works with stakeholders to design a coordination framework for the youth sector; however, this framework was not implemented by MoYA and NAYCOM due to the Ebola epidemic and, to some extent, tensions between the leadership of these institutions. Although the Coordination Mechanism Framework was supported by sensitization and awareness-raising programmes implemented among youth at district and chiefdom levels, the lead governmental youth institutions began to experience poor coordination due to the lack of implementation of the framework. This problem contributed to the fragmented implementation of the youth programmes by the different development entities.

4.2.13.5 Strategies Implemented to Resolve Organizational Challenges

The MoYA and NAYCOM collaborate with UNDP and other stakeholders to assess the institutions' challenges and develop an advocacy programme that promotes and encourages youth activism and participation in national development planning processes. The programme encourages communities to organize themselves, participate and develop need-based projects/programmes across the country at the national, district and chiefdom levels. The UNDP spearheaded work on the development of the Youth Coordination Mechanism Framework to enhance the levels and quality of programme management and coordination within MoYA and NAYCOM.

It was also used to mainstream youth development in the strategic programmes by governmental ministries and divisions across the civil service implemented. Youth mainstreaming is defined as:

> The process of assessing the implications for young persons, of any planned action, including legislation, policies, programmes, and projects, in all areas and levels. (United Nations Economic & Social Council (ECOSOC) in University of the West Indies, 2011; https://vdocuments.net/the-national-youth-mainstreaming-strategy-why-youth-mainstreaming-youth-are-both.htm)

The youth mainstreaming strategy was utilized by the MoYA and NAYCOM as a comprehensive strategy to formalize the concerns and experiences of young people in the national development planning process. Youth are viewed within the model as part of the solution and not merely passive objects of intervention. The model is part of an ongoing and wider initiative to strengthen the institutional capacity of the MoYA and NAYCOM to streamline and fortify youth development programme initiatives across the three sectors.

4.2.13.6 The Capacity Development Problems (Human) Facing the Organization

NAYCOM and the MoYA were confronted with capacity-building challenges that retard the development, management and implementation of the youth development programmes.

> The organisation also faced problems regarding data management, analysis for reporting, and contract management, theorized Mumudo Wudie, programme manager of NAYCOM.

There were technical deficiencies in both NAYCOM and NAYCOM regarding project implementation and accountability. The institutions were fraught with (a) weak and/or lack of accountability measures to ensure effective development and implementation of projects; (b) weak and/or lack of programme development, monitoring and evaluation reporting systems; (c) weak and/or lack of strategic direction and vision; and (d) weak and/or lack of strategy formulation and implementation of policies. These development challenges make it extremely difficult to manage governmental institutions to accomplish the programming objectives. Besides these shortcomings, some senior officials, such as Anthony Kamara, and youth commissioners were very committed and dedicated and worked diligently with UNDP to ensure that the projects implemented by NAYCOM were effectively and efficiently implemented. These officials worked nonstop to persuade policymakers and officials of the UNDP to continue supporting the YEEP.

4.2.13.7 Strategies Implemented to Solve the Human Capacity Development Problems

Due to the overwhelming challenges facing the Government of Sierra Leone regarding the low levels of human capacity in MoYA and the NAYCOM, it requested technical and financial resources from the UNDP to support the institutional strengthening of the capacity of NOYA and NAYCOM to enable these youth development agencies to effectively manage and coordinate their youth development programmes/projects. The UNDP assisted the Government of Sierra Leone in developing a Youth Coordination Development Framework, which provides a system for coordinating all programmes/projects across governmental institutions and beyond.

The framework was supported by developing a monitoring, evaluation and reporting (MER) system, which provides feedback to the policymakers and senior officials on the programme's performance. However, the Ebola pandemic impedes the full implementation of the project.

The YEEP provides strategic directions to the governmental ministries on the specific capacity development training, internship, coaching, networking and other initiatives to be implemented to achieve the government's goal. Through the project, the government was advised on the policies and roadmaps for the different development interventions it wishes to implement to build the resilience of young people and their communities to deal with the country's poverty and underdevelopment challenges. The capacity development programmes were geared towards engaging different categories or sectors of the population, including staff of MoYA and NAYCOM and other governmental ministries, representatives of civil society organizations involved in youth development, senior management and instructors of universities and colleges and TVET institutions, local government authorities and youth people and their organization including former child soldiers and other disadvantage and marginalized young people.

4.2.13.8 The Methodologies and Their Effectiveness Used to Build Capacity

Due to the intensity of the capacity development problems confronting MoYA and NAYCOM, a series of training workshops were implemented to enhance the knowledge and skills of government and civil society officials in new innovations in youth development practices, including governance, poverty alleviation, decent jobs and entrepreneurship development. The participants in these training workshops provided reference materials on various topics. Some senior staff was sent to regional and international training workshops and conferences where they acquired new skills and knowledge on development issues.

Unemployed youth who did not gain employment after completing their degrees were recruited in the Graduate Internship Programmes (GIP). They were provided with a stipend to cover their transportation and other cost and participated in 6-month programmes in an approved accredited institution. Upon completing their internship or apprentice programmes with the governmental and private sector institution, the young person would acquire skills and knowledge from the professionals they were assigned to boost their employability. The training courses and workshops would utilize methodologies such as whole-class presentations, panel discussions, peer-to-peer learning and group work exercises to stimulate and sensitize young people on critical issues by transferring new skills and knowledge.

There were differences in the views expressed by informants regarding the effectiveness of the methods used to enhance the skills and knowledge of young people:

> Not very effective because people will tend to revert to the usual routine of doing business since they considered such capacity developments as only personal capacity developments. It is not viewed as institutional level capacity-building; they should use to enhance the capacity within their institution where they are based in. Dr Molla McKennon, project manager, UNDP

While Mumudo Wudie, programme manager of NAYCOM, has an opposite view on the effectiveness of the capacity-building programmes:

> The methods used were very effective. The facilitators ensured consistency of approach and methodology. The coaching and follow-up methods built-in helped to sustain the gains of the capacity-building programs.

This view was also supported by Marbey Sartie, former programme analyst of YEEP, UNDP:

> The project has injected transformative inputs into the psyche of youth that will stay with them even after the end of the project funding.

Although there were differences in the views of the informants regarding the effectiveness of the methodologies implemented by the YEEP project team, based on the responses from informants and beneficiaries, one can conclude that learning new skills and knowledge was an important element of the YEEP outcomes, and it was achieved. Through the YEEP, the UNDP project manager encourages the officials of MoYA and NAYCOM to create the enabling mechanisms within their organization for continuously learning to take place by enhancing the practices, processes and strategies within the organization. These officials are encouraged to implement mechanisms to create a functional learning organization.

Learning organizations are where people continually expand their capacity to create the results they truly desire, where new and expansive patterns of thinking are nurtured, where collective aspirations are set free and where people are continually learning how to learn together (Senge, 1990, p. 1). The benefits derived from the YEEP are as follows:

- Increase the organization's vision and communication of the vision to individuals and other organizations.
- Top government officials' support and commitment to the organization are supported by bottom-up and top-down strategy implementation.
- Working on change quickly at all levels of the organization.
- Ensuring that the organizational systems support the desired changes anticipated with appropriate strategies.
- Leadership reinforcement of new behaviours through modelling.
- The creation of organizational systems encourages the regeneration of ideas and eliminates any notion of non-participation.
- Employee-assisted learning programmes and policies that encourage learning for its own value.
- Shift away from the competition with others towards continuous personal challenge.

It is impossible to identify the specific stage the organizations were at in implementing these learning strategies. Still, it is fair to say that the management was implementing several learning strategies within the organization and influencing the partners they were collaborating with to implement the learning strategies when the devastating Ebola epidemic struck the country in 2014–2015. The epidemic destroyed the capacity of both organizations to sustain the benefits of implementing the YEEP. Despite such challenges, the management could extract the lessons learned and best practices of the YEEP and is making concerted efforts to continuously implement these lessons at all levels of the organization.

One of the critical lessons evolving from the YEEP is that the young people and their parents in Sierra Leone would place greater value on the training provided through a formal institution, resulting in the award of accredited certification to the individual after successfully completing such training as alluding to by Dr. Molla McKennon, project manager, UNDP. There is also a tendency for government and private sector institutions to view the knowledge and skills provided through programmes such as those offered by the UNDP YEEP and others as personal and of little value. However, the training provided is more relevant to the country's development needs, and the trainers are much more qualified than those from the tertiary intuitions that normally provide such accredited training.

The UNDP must work with the Government of Sierra Leone to put a mechanism in place to value the training provided to young people through the YEEP and other donor-funded skills development training programmes. UNDP and NAYCOM must work with the universities, colleges and TVET centres and the NCTVA to upgrade the tertiary skill education curriculum, the training of instructors in the use of innovative training methods and technologies supported by the provision of new tools and equipment to these institutions as a strategy for enhancing the quality of training and linking the training to the needs of the market.

4.2.13.9 The Key Lessons Learned from Implementing the Capacity Development Programmes

Capacity development works must be formulated and packaged successfully to be implanted in institutions. The institutions should implement mechanisms to retain their capacities when persons depart from the organization. The capacity development programmes implemented through the YEEP not only enhance the skills and knowledge of youth to gain sustainable livelihoods but, most importantly, contribute to harnessing the leadership and management capacity of MOY and NAYCOM to champion the need for investing in the capacity organization. The capacity-building ambitions utilized by UNDP YEEP matched the different realities of staff, youth and partner institutions in Sierra Leone.

The lessons learned from YEEP are:

- YEEP has contributed to enhancing the capacity of its implementing partners. The staff of these agencies were involved in the training and consultations process and acquired new skills and competencies which contributed to the overall management and operation of the youth development programmes and the partner institutions.
- The improvement of the technical capacity of the institution, specifically their monitoring, reporting, training, reporting and tracking, contributed to enhancing the overall management of the youth development programme in Sierra Leone.
- Beneficiaries of the YEEP utilized the skills acquired from the YEEP to deliver services to stop the Ebola epidemic, including (a) contact tracers, (b) coordinators, (c) communication and (d) data management clerks and others.
- The members of staff of the NAYCOM and MoYA and partner agencies were motivated and passionate about working with young people to enhance their own employment situation and that of their community, thus contributing to the achievement of national sustainable development goals.
- The BDS centres provide training to participants to enhance their capacity and leadership skills, which boosts their business and employment skills and leads to the development of their own businesses. These businesses employed a considerable amount of young people in participating communities (an average of five youth per business).
- The mainstreaming of sustainable youth development criteria in the programmes of the BDS, CAPS and GIP centres.
- Students graduating from universities and colleges are using the advice provided by CAPS to gain employment with government and private sector institutions across Sierra Leone.
- Coordinators at the CAPS centre of participating colleges and universities created a WhatsApp group to facilitate the sharing of experiences, best practices and advice from colleagues on strategies for taking the CAPS forward.
- The provision of computers and internet facilities to partner colleges and universities makes students computer literate and aids communication and knowledge acquisition.
- YEEP contributed to building the confidence, independence and creativity of beneficiaries. According to one beneficiary: "I am in a position where I can pay my rent and take care of the members of my family. I have a bank account now where I can put an amount of savings aside".
- YEEP aided in increasing beneficiaries' understanding of businesses and provided support to enhance young people's participation in youth and community groups, including football and other sporting organizations. The YEEP indirectly utilized sports as a tool for the achievement of sustainable development goals.
- Young people can now pay their taxes and city rates, thus strengthening the national development processes.

4.2.13.10 The Best Practices

Human capacity development programmes must be based on the needs of the institutions and should be implemented to enhance the institution's capacity. These programmes must be customized and oriented to improve the visibility of the institutions as well as strengthen collaboration, networks and partnerships with partner agencies. Therefore, the effective coordination of programmes is fundamental for successful youth development programmes' successful management and outcomes. The best practices evolving from the YEEP are:

- Young people are enthusiastic and possess the zeal, passion and aptitude to pursue their own development. However, agencies such as

NAYCOM and MoYA must collaborate with UNDP and other international development organizations to bring key partners on board to provide resources and technical, strategic advice to drive youth development programmes.

- An increase in the number of persons desirous of accessing the BDS programme since the values placed on the quality of the training programme by governmental institutions, businesses and development agencies alike.
- Provision of incentives, such as transportation, lunch, etc., to marginalized young men and women by the YEEP management to facilitate that poor young men and women benefit from the project who would not usually participate in such tasks due to their low poverty status.
- Young people are motivated because they are determined to improve their livelihood and are entreating providers of adequate opportunities. This is demonstrated by young people spending over Le 70,000 (US$7) per day to cover transportation, food and other costs. This is shown by the high attendance rate of participants at the training centres delivering these programmes (85–90% daily attendance rate).
- Students in universities and colleges and TEVT centres make tremendous efforts to choose appropriate careers aligned with their interests and competencies because of the advice provided through the CAPS initiative.
- CAPS coordinators and beneficiaries are motivated to make greater efforts to ensure the programme succeeds.
- Effective time management is an important consideration to ensure the employment of GIP and CAPs interns.

4.2.13.11 The Recommendations for Improving Capacity-Building

The capacity-building interventions should be done to enhance the operational mechanism and governance processes within an institution by creating the enabling environment for the institution to impact positively on young people and the achievement of national and sustainable development goals:

- The development and implementation of a sound staff appraisal system backed up with opportunities for the promotion of staff, including identifying employees that work hard and work well, appreciating their efforts openly so that they feel recognized and then rewarding them with the right incentives to keep their enthusiasm going.
- The National Youth Commission is a collection of human beings working together, so building core relationship skills, like the ability to collaborate and communicate, is one of the most important things that NAYCOM can encourage.
- Encourage a pro-partnering culture by assigning staff the time to develop partnerships and give them the freedom to be creative and innovate, including the freedom to fail. Reward staff for the building of relationships, as well as the results that come from fully formed partnerships.
- There is a need to up-scale the components of YEEP by expanding and roll-out the element of the programmes nationally, specifically in vulnerable and marginalized communities.
- Enhance the coordination and partnership among youth development organizations and stakeholders to augment the management, funding and reporting of YEEP.
- Facilitate the mainstreaming of YEEP and broader youth sustainable development issues across government agencies and with partners at the national, district and chiefdom levels.
- Make YEEP more responsive and flexible to the needs of young people and the community.
- Mobilize donor financing by creating a youth development trust fund to ensure the availability of sustainable finance to support the development and expansion of the YEEP and other youth development initiatives.
- Explore the possibilities of mobilizing financial resources to enable the YEEP to provide bus services to BDS centres to reduce the cost of transportation of beneficiaries and volun-

teers attending training (currently, it costs participants an average of Le 75,000 per day to attend training at the BDS centres).

- Expanding the time for GIP from 6 months to 1 year to provide more time for young people to acquire knowledge, skills and competencies in the workplace.
- Due to the magnitude of social and economic challenges facing young people in Sierra Leone, the CAPS programme must focus on providing social counselling services to complement the academic counselling service currently provided to students.
- Mainstreaming the BDS initiative into the livelihood development programme of the government youth development initiative.
- Provision of more financial resources from government and donors to support business start-ups, livelihoods initiatives and the development of business incubation centres to provide space for new business start-ups.
- Explore strategies for making the BDS, GIP and CAPS centres more sustainable through the government allocating more local resources to support the programme creation of a mechanism to enhance the participation of the community in the management and operation of the YEEP programme. NAYCOM should mobilize more development partners and private sector agencies to provide technical and financial resources to support the YEEP.
- On completion of CAPS, GIPS and BDS studies, each trainee should be given a laptop and/or similar device to motivate them to become computer literate.
- Increase the number of interns from 300 to 1000 participants to have a greater impact on the country's economy.
- Showcasing of GIP and other youth development programmes as a strategy for motivating young people to participate in the YEEP and similar youth development initiatives.
- Increase in young women interns through direct targeting utilizing gender mainstreaming as a strategy for attracting more women to the YEEP programme. Therefore, an empowered young woman would contribute to an empowered family and an empowered community.
- Sharing the outcomes of the YEEP with relevant development partners and showcasing the achievements of the beneficiaries of the YEEP through skills employment fair, TVET promotion campaigns and other events.
- Intensification of public education mechanism on the YEEP, including (a) human development and career training for interns, (b) effective media engagement and documentation of YEEP, (c) implementation of further research on the issues facing youth empowerment and employment and (d) restructuring the YEEP to make it more relevant to the development needs of the country.
- Enhance resources to support the activities and programme of partners to effectively deliver the YEEP in communities away from Freetown, the capital.
- Promoting trade fairs and career days provides opportunities for YEEP beneficiaries to showcase their achievements, creativity, innovation and business acumen.

References

Angelsen, A., Larsen, H. O., Lund, J. F., Smith-Hall, C., & Wunder, S. (2011). *Measuring livelihoods and environmental dependence; methods for research and fieldwork*. Earthscan.

Armstrong, M. (1999). *How to be an even better manager* (5th ed.). Kogan Page Limited.

Becker, A., & Kretsch, E. (2019). The leadership void for climate adaptation planning: Case study of the Port of Providence (Rhode Island, United States). *Frontiers in Earth Science, 7*, 29. https://doi.org/10.3389/feart.2019.00029

Burnes, B. (2000). *Managing change – A strategic approach to organisational dynamics* (3rd ed.). Pearson Education Limited/Prentice Hall Publishing.

Butler, J. R. A., Wise, R. M., Skewes, T. D., Bohensky, E. L., Peterson, N., Suadnya, W., Yanuartati, Y., Handayani, T., Habibi, P., Puspadi, K., Bou, N., Vaghelo, D., & Rochester, W. (2015). Integrating top-down and bottom-up adaptation planning to build adaptive capacity: A structured learning approach. *Coastal Management, 43*, 346–364. Taylor & Francis Group, LLC; ISSN: 0892-0753 print/1521-0421 online. https://doi.org/10.1080/08920753.2015.1046802

Chambers, R. (2002). *Participatory workshops: A sourcebook of 21 sets of ideas and activities* (Vol. 794, pp. 421–434). Earthscan.

Chanika-Mataya, D., Vincent, K., & Dougill, A. J. (2020). How can we effectively build capacity to adapt to climate change? Insights from Malawi. *Climate and Development*. Francis and Taylor Group, *12*(9), 781–790. https://doi.org/10.1080/17565529.2019.1694480

Charles, H. (2006). *The Commonwealth's response to global challenges facing youth in the 21 century* (Vol. 4, No. 2). Commonwealth Youth and Development, Department of Commonwealth Studies, UNISA (University of South Africa Press).

Commodore, A., Wilson, S., Muhammad, O., Svendsen, E., & Pearce, J. (2017). Community-based participatory research for the study of air pollution: A review of motivations, approaches, and outcomes. *Environmental Monitoring and Assessment, 189*, Art. 378.

Commonwealth Secretariat. (2007). *The commonwealth plan of action on youth empowerment 2007–2015*. Commonwealth Youth Programme, Commonwealth Secretariat.

Danielsen, F., et al. (2009). Local participation in natural resource monitoring: A characterization of approaches. *Conservation Biology, 23*, 31–42.

Eicken, H., Danielsen, F., Sam, J. M., Fidel, M., Johnson, N., Poulsen, M. K., Lee, O. A., Spellman, K. V., Iversen, L., Pulsifer, P., & Enghoff, M. (2021). Connecting top-down and bottom-Up approaches in environmental Observing, *BioScience, 71*(5) 467–483.

Feiner, M. C. (2002, November 15). Laws of leadership, mastering leadership. *Financial Times*.

Government of Sierra Leone. (2010). *The second joint progress report on the agenda for change June 2010 – June 2011, Government of the Republic of Sierra Leone, Freetown, Sierra Leone*. https://devinit.org/924633#c6ee5bea (Development initiatives based on World Bank PovcalNet and IMF World Bank Economic Outlook).

Government of Sierra Leone. (2013). *The agenda for prosperity road to middle-income status: Sierra Leone's Third Generation Poverty Reduction Strategy Paper (2013–2018): Freetown, Sierra Leone*. http://faolex.fao.org/docs/pdf/sie149110.pdf

Government of Sierra Leone. (2015). *Sierra Leone 2015 population and housing census thematic report on population structure and population distribution; Freetown, Sierra Leone.*

Government of Sierra Leone. (2020, December). *National Youth Policy 2020–2025: Empowered youth leading the development of a new Sierra Leone, Freetown, Sierra Leone.*

GYNED-SL. (2020). *Building the institutional capacity of Youth Community Based Organisations (YCBO) to combat desertification and land degradation in Sierra Leone, Freetown, Sierra Leone* (Unpublished doc).

Hernandez-Delgado, E. A. (2015). *The emerging threats of climate change on tropical coastal ecosystem service, public health, local economies, and livelihoods sustainability of small islands: Cumulative Impacts and Synergies*. Marine Pollution Bulletin, University of Puerto Rico, Centre for Applied Tropical Ecology and Conservation, Coral Reef Research Group Puerto Rico, USA.

Horstmann, B., & Schulz-Heiss, G. (2014). *Providing international adaptation finance for vulnerable communities: A study on potentials and limits of social investment funds*. The German Development Institute. ISSN 1860-0468.

https://sierraleone.unfpa.org/en/node/6136

https://www.oecd.org/dac/evaluation/daccriteriaforevaluatingdevelopmentassistance.htm

https://www.sl.undp.org/content/sierraleone/en/home/about-us.html#:~:text=UNDP%20supports%20the%20people%20of,and%20democratic%20insitutions%20and%20processess

https://www.theconversationspace.com/products/listen-down-2/?src=g; https://www.undp.org/

ILO. (2010, November). *A skilled workforce for sustainable growth: A G20 training strategy*. International Labour Office.

ILO. (2013). Sustainable development, decent work and green jobs. In *International Labour Conference, 102nd session*. International Labour Office.

IPCC. (2014). Climate change: Synthesis report. In *Contribution of Working Group 1, 11, 111 to the Fifth Assessment Report of the Intergovernmental Panel on Climate Change* (p. 151). IPCC.

IPCC. (2022). *Climate change 2022 mitigation of climate change, summary for policy makers: Working Group 111 Contribution to the Sixth Assessment Report of the International Panel on Climate Change (AR6 WG 111)*. IPCC.

Johnson, G., & Scholes, K. (2002). *Exploring corporate strategy* (6th ed.). Prentice Hall.

Kallman, M. E., & Clark, T. N. (2016). *The third sector: Community organisation, NGOs and Nonprofits*. University of Illinois Press.[1]

Matland, R. (1995). Synthesising the implementation literature: The ambiguity-conflict model of policy

---

- 1 *Key informants, UNDP/Government of Sierra Leone YEEP*Molla McKennon, Programme Manager, YEEP, UNDP SL
- Marbey Sartie, Programme Analyst, UNDP SL
- Mamadu Wudie, Programme Manager, NAYCOM

- *Key Informants, GYNED-SL*Musa Koroma, Executive Director, GYNED-SL
- Alvian B. K. Sesay, Communication Officer, GYNED-SL
- Moses Kalokoh Putin, Media Official, GYNED-SL
- Aminata Olimatu Kamara, Gender Officer
- Ostilda C. O. Hughes, Operations Director, GYNED-SL
- Alma M. E. N. Sheriff, Director of Programmes
- Hawa Turay, Admin Assistant

implementation. *Journal of Public Administration Research and Theory, 5*(2), 145–174.

Nurse, L. A., McLean, R. F., Agard, J., Briguglio, L. P., Duvat-Magnan, V., Pelesikoti, N., Tompkins, E., & Webb, A. (2014). Small islands. In V. R. Barros, C. B. Field, D. J. Dokken, M. D. Mastrandrea, K. J. Mach, T. E. Bilir, M. Chatterjee, K. L. Ebi, Y. O. Estrada, R. C. Genova, B. Girma, E. S. Kissel, A. N. Levy, S. MacCracken, P. R. Mastrandrea, & L. L. White (Eds.), *Climate change 2014: Impacts, adaptation, and vulnerability. Part B: Regional aspects* (Contribution of Working Group II to the Fifth Assessment Report of the Intergovernmental Panel on Climate Change) (pp. 1613–1654). Cambridge University Press.

OECD. (2013). *OECD skills outlook 2013. First results from the survey of adult skills*. OECD.

OECD DAC. (2019). *Better evaluation, revised evaluation criteria. Definition and principles for use*. OECD/DAC Network on Development Evaluation. Revised evaluation criteria Dec 2019. pdf.

Robson, C. (2002). *Real world research, a resource for social scientists and practitioner researchers*. Blackwell Publishers.

Saunders, M., Lewis, P., & Thornhill, A. (2000). *Research methods for business* (2nd ed.). Pearson Education Limited.

Senge, P. M. (1990). *The fifth discipline*. Doubleday.

Simmons, A. (2006). *The commonwealth youth development index: Challenges and prospects for youth* (Vol. 4, No. 2). Commonwealth Youth and Development, Department of Commonwealth Studies, UNISA (University of South Africa Press).

Simmons, A. (2018). *Impact of climate change on youth in small island communities: The case of St. Vincent and the Grenadines*. This thesis is submitted to De Montfort University in partial fulfilment of the requirements for the award of Doctor of Philosophy: School of Engineering and Sustainable Development De Montfort University, Leicester, UK.

Simmons, A. (2021). *Impact of climate change on youth in small Island developing communities* (Sustainable development series). Springer.

Simmons, A. (2022). *Climate change adaptation framework and youth entrepreneurship in West Africa* (Sustainable development goals series). Springer.

UNDP. (2015). *Human development report*. Work for Human Development, United National Development Programme (UNDP).

UNESCO. (1978). *Records of the general conference. 20th session* (Vol. 1). UNESCO.

University of the West Indies. (2011). *The national youth mainstreaming strategy workshop: Presentation of strategy document June 29–30, 2011*. The Centre for Leadership and Governance, The University of the West Indies, The Ministry of Youth, Sports & Culture, The National Centre for Youth Development and the International Development Bank. https://vdocuments.net/the-national-youth-mainstreaming-strategy-why-youth-mainstreaming-youth-are-both.htm

Vagvolgyi, R., Coldea, A., Dresler, T., Schrader, J., & Christoph-Nuerk, H. (2016). A review about functional illiteracy: Definition, cognitive, linguistic, and numerical aspects. *Frontiers in Psychology*. https://doi.org/10.3389/fpsyg.2016.01617

5 Capacity-Building at the International Level: Stories from the Field

Abstract

This chapter presents case studies on the Commonwealth Youth Programme (CYP), Caribbean Youth Environment Network (CYEN) and Recovery of Acholi Youth (RAY) project in Northern Uganda work undertaken to build the capacity of young people and their institutions by enhancing their resilience to tackle the development challenges in their community and across the Commonwealth. The case studies explore the strength, weaknesses/challenges, methodologies implemented and their effectiveness, the lesson learned and best practices and present recommendations for enhancing the capacity of the institutions now and into the future.

The first case study presents an analysis of the work of the CYP. The CYP is the Youth Affairs Division (YAD) of the Commonwealth Secretariat based in London, UK. The institution works with governments and youth NGOs in 54 countries in Europe, Africa, Asia, the Caribbean and the Pacific to build their institutional capacity by utilizing top-down approaches to support good governance, peace, the rule of law and the SDGs in member states of the Commonwealth.

The second case examines the work of the CYEN's capacity development initiatives implemented in 15 countries in the Caribbean region. CYEN is a regional youth environment and development network established by young people 30 years ago and is based in Barbados. The third section of the chapter explores the capacity-building strategies implemented by the Norwegian Refugee Service (NRS) in the RAY project in Northern Uganda. Through international cooperation and partnership, the RAY involved several developmental organizations utilizing education and sustainable livelihood strategies working in collaboration to address the refugee crisis brought about by the civil war in Northern Uganda.

Keywords

Commonwealth Youth Programme · Caribbean Youth Environment Network · Acholi youth · Northern Uganda · Development challenges · Community · Education · Commonwealth · Non-governmental organization · Refugees

This chapter presents case studies on the Commonwealth Youth Programme (CYP), Caribbean Youth Environment Network (CYEN) and Recovery of Acholi Youth (RAY) project in Northern Uganda to build the capacity of young people by enhancing their resilience to tackle the development challenges in their community,

A. Simmons, A.-A. Simmons, *Building Institutions for Resilience*, Sustainable Development Goals Series, https://doi.org/10.1007/978-3-031-28881-4_5

country and regions. The case studies explore the strength, weaknesses/challenges, methodologies implemented and their effectiveness, the lesson learned, the best practices and present recommendations for enhancing the capacity of the institutions in the future.

The first case study presents an analysis of the work of the CYP. The CYP is the Youth Affairs Division of the Commonwealth Secretariat based in London, UK. The institution works with governments and youth NGOs in 54 countries in Europe, Africa, Asia, the Americas, the Caribbean and the Pacific regions to build the institutional capacity utilizing top-down approaches to support good governance, peace, the rule of law and SDGs in member states (OECD, 2013; Matland, 1995). The second case examines the work of the CYEN's capacity development initiatives implemented in 15 Caribbean countries in the Caribbean region. CYEN is a regional youth environment and development network established by young people 30 years ago and is based in Barbados. The third section of the chapter explores the capacity-building strategies implemented by the Norwegian Refugee Service in the RAY project in Northern Uganda. The original project was started by the Commonwealth Youth Programme and involved several developmental organizations working to resolve the refugee crisis brought about by the civil war in Northern Uganda through international cooperation and partnership.

5.1 Case Study on the Commonwealth Youth Programme: Development of Participatory Integrated Development Model to Enhance Capacity-Building and Sustainable Development Goals (SDGs) Among Youth in Member Countries of the Commonwealth (1997–2006)

The Commonwealth of Nations is a voluntary association of free nations that grew out of the former British Empire. Growing from a few former colonies, such as India, that gained their independence from Britain after the Second World War, the association grew larger in the 1950s and 1960s as more African, Asia and Caribbean colonies gained independence. There are currently 54 member states that make up the organization. The number was extended to include Namibia in 1990, South Africa in 1994, Mozambique in 1995 and Rwanda in 2009. These countries decided to remain together as an association of free nations. Queen Elizabeth 11th of England was the head of 15 of the 54 member states. In addition to their common historical experiences of British rule, the Commonwealth has also remained bound together by common traditions and legacies of common law, systems of education, public administration, culture and the common use of the English language in their political, diplomatic and economic and trade relations (Takawira & Simmons, 2004; Simmons, 2006; https://thecommonwealth.org/about-us).

The Commonwealth of Nations consisted of 54 independent countries, both advanced and developing economies located in Africa, Asia, Europe, the Caribbean and the Pacific regions. The Commonwealth has a population of about 2.5 billion people out of the 7.9 billion globally (https://www.Commonwealth.org), of which 60% are under the age of 29. Approximately two-thirds of members of the Commonwealth are classified as small states. Of that amount, thirty-two (32) of the world's forty-two (42) small states are Commonwealth members (population of 1.5 million people). Twenty-five countries which are members of the Commonwealth are classified as small island states. Many of the Small Island Developing States (SIDS) are recognized as a distinct, diverse group of developing countries encountering specific social, economic and environmental vulnerabilities at the United Nations Conference on Environment and Development (UNCED in Rio in 1992), also known as the Earth Summit, held in Rio de Janeiro, Brazil (3–14 June 1992) (https://www.un.org/ohrlls/content/about-small-island-developing-states#:~:text=Small%20Island%20Developing%20States%20(SIDS,social%2C%20economic%20and%20environmental%20vulnerabilities).

SIDS are particularly vulnerable to climate change and its effects. Sea-level rise (SLR), changing precipitation and rainfall regimes, changes in tropical storms and cyclone frequency or intensity, increasing air and sea surface temperatures, changes in atmospheric chemistry and melting of mountain glaciers are some of the phenomena and disruptions related to climate change (Gomez, 2013; IPCC, 2014a, b, c, 2022). SIDS are by no means a homogenous group of countries; they differ in terms of geography, physical, climate, social, political, cultural and ethnic characteristics and by stages of economic development. Besides these differences, the islands tend to share several characteristics that not only identify them as a distinct group but also underscore their vulnerability in the context of sustainable development and climate change (Nurse et al., 2014; IPCC, 2014a, b, c). These include (a) limited physical size, which effectively reduces some adaptation options to climate change and SLR; (b) generally limited availability of natural resources, i.e. many of the resources are already under severe stress from unsustainable use by human activities; (c) relatively thin water lenses that are highly sensitive to SLR, generally high population densities and in some cases high population growth; and (d) a poorly developed infrastructure (UNEP, 2013). However, these development challenges are compounded by the limited financial and human development skills, which can severely limit the capacity of small islands to adapt to climate change from a global economic perspective (Simmons, 2018).

Despite the above development challenges facing Commonwealth member countries, many of the countries are among those with the lowest ecologic footprint worldwide. Yet, Small Island Developing States (SIDS) are the most vulnerable and severely impacted by climate change and other development challenges (https://thecommonwealth.org/about-us).

The ecological footprint is a method promoted by the Global Footprint Network to measure human demands on natural capital, i.e. the quantity of nature it takes to support people or an economy. It tracks this demand through an ecological accounting system ().www.ecologicalfootprint.org

5.1.1 The Situation of Young People in the Commonwealth

The situation of young people globally is a challenging one. The global youth landscape indicates that young people, specifically those from developing countries, are more vulnerable than their counterparts in developed countries (UNDP, 2015). Issues and challenges such as poverty, discrimination and inequality and lack of access to proper education and health services continue to dominate their lives (Simmons, 2006; UNDP, 2015). Of the 2.5 billion people living in the Commonwealth, over 60% of the population are young. Young people are the most vulnerable sector of the population due to the severity of the impact of climate change. Climate change impacts youth and their communities, enterprises and overall well-being, specifically in locations exposed to storms, floods, droughts and fires (ILO, 2011).

The lack of opportunities for youth is a dismal calamity for Commonwealth SIDS. This problem is compounded by an increase in migration, especially among the most educated youth, resulting from youth being unable to visualize a future within their Commonwealth developing states after not finding employment. They leave their countries leading to a brain drain, which saps the country's sustainable future on top of the severe impact countries are already experiencing due to climate change (UNEP, 2013). Youth in Commonwealth SIDS and least developed countries (LDC) also suffer from reduced formal education and training prospects within their communities.

Unemployment and underemployment are the most serious issue facing Commonwealth countries. The average youth unemployment rate in the Commonwealth countries is 22.9%, compared to the global average of 19.2% (www.yourcommonwealth.org). The unemployment problem is caused mainly due to the absence of appropriate skills which would enable young people to be meaningfully employed (UNDP, 2015; ILO, 2010). Over the years, despite marked improvement in living standards and education in Commonwealth countries, young people are still largely isolated from participation in the mainstream economies of their various countries, both in terms of actual economic activities in influencing policies and programmes that affect their participation. Important implications of this "disenfranchisement" include increasing levels of poverty among youth, increasing instances of youth suicide; the acute rise in youth involvement in the use of, and trade in, illicit drugs; increasing numbers of young people committed to prisons and mental institutions; and the sharp increase in the prevalence of sexually transmitted diseases among young persons (Charles, 2006; Simmons, 2006; Carter, 2008).

Charles (2006, p. 49) describes the unemployment situation in the Commonwealth as follows:

> A plethora of youth enterprises programmes and diverse models for youth employment schemes have been implemented…unfortunately, many of these are underfunded and not adequately aligned to national development or to sustainable livelihoods policies and goals. To compound the problem, there is seldom a credible mechanism that allows for evaluation of the impact of these initiatives on the lives of the most vulnerable and marginalised youth. It becomes difficult to determine whether these significantly affect youth employment and economic empowerment.

Unemployment and underemployment are the most serious issues affecting the Commonwealth's young people. Young people generally account for 41% of the world's unemployed. However, this figure can be as high as 50–70% in some developing countries of the Commonwealth. Statistics have shown that over 208 million (18%) of the 1.09 billion affected young people globally (15–24 years old) survive on less than US $1 (one dollar) per day and 515.1 million (45%) on less than US $2 (two dollars) per day (UNDP, 2015).

In Commonwealth countries, the average youth unemployment rate is 22.9%, compared to the global average of 19.2% (www.yourcommonwealth.org). A significant challenge facing youth in the Commonwealth is the absence of appropriate skills which would make them meaningfully employed (UNDP, 2015; ILO, 2010). Over the years, despite marked improvement in living standards and education in Commonwealth countries, young people are still primarily isolated from participation in the mainstream economies of their various countries, both in terms of actual economic activity and in influencing policies and programmes that affect their participation. Important implications of this "disenfranchisement" include increasing levels of poverty among youth, increasing instances of youth suicide; the acute rise in youth involvement in the use of, and trade in, illicit drugs; increasing numbers of young people committed to prisons and mental institutions; and the sharp increase in the prevalence of sexually transmitted diseases among young persons. Young people are the greatest resource for development in the future of the Commonwealth of nations, yet they are among the most vulnerable of the social groupings. Therefore, young people must be empowered to control their lives and positively transform their societies.

5.1.2 The Commonwealth Youth Programme

The Commonwealth Youth Programme (CYP) is the Youth Affairs Division of the Commonwealth Secretariat that plans, coordinates, implements and evaluates all youth development programmes in the Commonwealth (www.yourcommonwealth.org). It was established by the Commonwealth heads of government at their meeting in 1973 as a response to the growing need of young people in the Commonwealth member countries. The Commonwealth Youth Programme forms a core and integral part of the Commonwealth Secretariat

Strategic and Operational Plan and programme of activities. The Commonwealth Youth Programme is also unique in that it is financed by a dedicated fund from voluntary contributions by all the member states of the Commonwealth. The CYP also receives funding from other non-governmental sources, collaborating partners and developing agencies.

The CYP operates in the 54 member countries of the Commonwealth. It works directly with governments, young people and other youth development regional and international agencies working in specific countries of the Commonwealth.

The programmes of the CYP are premised and embedded in a strategy that promotes the development of Commonwealth fundamental principles and values of democracy, good governance, respect for human rights, the rule of law and gender equality. The Commonwealth also works with other agencies, such as the United Nations, on major global issues such as conflict prevention, world trade issues, HIV/AIDS and the attainment of the Global Development Goals, formally the Millennium Development Goals. As an integral part of the Commonwealth Secretariat, the CYP provides technical support to member governments through Ministries of Youth, the National Youth Councils (NYC) and youth development networks to support these Commonwealth global development efforts (Commonwealth Secretariat, 2007).

5.1.2.1 The Uniqueness of the CYP Regional Centre

The CYP was the only division of the Commonwealth Secretariat with a regional presence. The CYP is financed by a dedicated fund from voluntary contributions from member governments of the Commonwealth. It also receives funding from the development partners it is working in partnership with a diverse range of collaborative youth development activities. Regional centres were established in Asia (India), Africa (Zambia), Caribbean (Guyana) and the Pacific (Salmon Island) regions as regional hubs to coordinate the delivery of its programmes to member countries. The strategic programmes are managed through the Youth Affairs Division (YAD) based at the Commonwealth Secretariat in London. Although these regional centres have since closed, they provided a regional presence and supported the delivery of worthwhile and valuable programmes to the regions.

> provides support and services to Ministries of Youth and youth development networks across the . The Centre offered programmes based on 's 'global' Strategic Plan in 4 components: (a) Youth Work; (b) Youth Response to HIV/AIDS; (c) Youth Policy, Strengthening Youth Ministries and Youth Networks; and (d) Youth Micro-Entrepreneurship. Several annual milestone events emanated from or were conjoined with this programming, for example, the Youth Ministers' Meeting, Youth Day, the Youth Exchange, and the establishment of the Regional Youth Council. Competency Standards for Youth Work were developed, transitioning into the undergraduate degree course in Youth Work and Development, currently offered by the University of the West Indies. Under component (d), several micro-entrepreneurship programmes established and supported through building local capacity in line with the Youth Credit Initiative (CYCI) model. Dwynette Eversley, former programme manager, Youth Programme.

5.1.2.2 Policy and Governance Approaches Instituted by CYP

Before 1997, CYP implemented a top-down approach to developing and implementing its youth development programmes. The organization experienced difficulties delivering its programmes to young people in member countries. Youth were engaged in developing and delivering the youth programmes due to the rigid top-down nature of the approach instituted by the government and CYP. The agency implemented a hybrid approach to enhance the delivery of its programme and instituting management and governance structures to enhance the participation of young people in the organization's governance. These governance structures include the participation of youth in the Committee of Management (COM) of the Commonwealth Youth Programme and the creation of the Commonwealth Youth Caucus (CYC).

The hybrid approach is a combination of the top-down and bottom-up approaches. Combining both approaches highlights the strengths and minimizes the weaknesses of the approaches. The hybrid approach was instituted when Jane Foster undertook the position of head of CYP and the Youth Affairs Division at the Commonwealth Secretariat in London, UK (Matland, 1995; OECD, 2013). The change in the approach of the CYP was a result of member governments being unhappy with the direction and outcomes achieved thus far and calling for a change in the orientation of the programme and composition of the staff at the Pan-Commonwealth Office.

Policy implementation is most likely to be successful when all stakeholders at various levels can interact with each other. This provides adequate opportunities for central policymakers at the Pan-Commonwealth Office in London to work directly with government ministers and senior officials at the Ministries of Youth and the youth development networks and NGOs operating in member countries. The hybrid approach provides opportunities for local stakeholders, including youth and community leaders, to work with each other and directly with policymakers and senior officials in their respective countries. This situation contributed to the successful implementation of policy and strategic programmes in Commonwealth countries (OECD, 2013). While the successful outcomes of policy are determined by the efficient delivery of projects and programmes at the community level, the roles central government plays in implementing youth development programmes are critical to policy development. Hence centrally located government officials can exercise their influence by providing resources and incentives to local stakeholders, focusing attention on development issues, establishing strong leadership around a vision for the policy and engaging with networks of local groups and communities (Stachowiak et al., 2016). The hybrid approach was highly favoured by all stakeholders and was responsible for the successful implementation of CYP programmes.

The CYP governance and management mechanisms are structured to ensure that young people participate actively at all levels of decision-making. The Commonwealth Board of Governors, through its Executive Committee, is the main governance mechanism of the CYP. It provides the strategic direction of the CYP programmes. The Executive Committee is responsible for the formulation of the policies to guide the operation of the CYP. The board consists of the representative of the Commonwealth high commissioners based in London and the head of the Commonwealth Youth Caucus. The Youth Caucus is another structure of the CYP that derives its mandates and operates at national and regional levels (Takawira & Simmons, 2004).

The Executive Committee comprises 16 members elected by the Board of Governors of the Commonwealth. It is responsible for overseeing the implementation of the decisions of the Board of Governors by the Commonwealth Secretariat. The board meets quarterly, and young people are represented on the Board of Governors and the Executive Committee. During the period under review, CYP operates through its Regional Advisory Boards (RABS). The RABS are responsible for advising on the programme's operations within respective regions of the Commonwealth. The RABs comprised senior government representatives and youth from member countries. CYP provides technical support to member governments through the Ministry of Youth Affairs, the National Youth Councils and other youth development networks in Support of the Commonwealth development endeavours (Takawira & Simmons, 2004).

5.1.2.3 CYP Conceptual Framework on Youth Empowerment

CYP work is based on young people's genuine and active participation in the development process (Takawira & Simmons, 2004). This approach views development as an inseparable component of human rights, and the active participation of young people is fundamental to their development. Young people's involvement in planning, implementing, monitoring and evaluating programmes that affect them makes these programmes accountable (Commonwealth Secretariat, 2007). The rights-based approaches to youth development as applied to the participation of young people are well supported and advocated by the United Nations Convention on the Rights of the Child (UNCRC). A key component of the approach is young people should share in the burden and benefits of national development and responsible citizenship (Simmons, 2006; Takawira & Simmons, 2004). Youth participation in theory and practice underpins the work of CYP with young people, and the organization encourages member governments to adopt the same youth development paradigm.

5.1.2.4 Global Youth Development Index (GYDI)

CYP developed a framework of indicators to assist it in monitoring and evaluating the performance of the youth development programmes implemented by member governments. The 2020 Global Youth Development Index (GYDI) shows that the circumstances of young people have improved around the world by 3.1% between 2010 and 2018; however, the improvement remains slow (Commonwealth Secretariat, 2021). The index ranks countries between 0.00 (lowest) and 1.00 (highest) according to the developments in youth education, employment, health, equality and inclusion, peace and security and political and civic participation. CYP provides leadership and collaboration with international development partners to develop a framework of 27 indicators to show the state of the world to 1.8 billion young people between the ages of 15 and 29, including literacy and voting within the ages of 15–29. The GYDI is viewed as a framework that provides measurement to evaluate changes in the status of youth development at human, institutional and national levels. Overall, the GYDI focuses on measuring a few critical aspects of youth development that lend themselves to statistical and socio-analytical methods that can provide information on the status of youth (Simmons, 2006).

The GYDI provides policymakers with an important instrument that is useful in shaping the direction of youth development work and policies. The GYDI helps Youth Ministries and policymakers determine how to allocate scarce resources among competing priorities that affect youth development, thus reducing duplication of efforts and wastage of resources (Simmons, 2006; Commonwealth, 2021).

5.1.3 Aim and Objectives of CYP

The overall aim of the Commonwealth Youth Programme is to enable young people to realize their own potential in the Commonwealth to transform and improve the quality of their lives in partnership with other stakeholders in a way that they are recognized as valued stakeholders as part of their society (Takawira & Simmons, 2004).

The objectives of the CYP are as follows:

- Promote Youth Enterprise Development and sustainable livelihood in Commonwealth member countries to alleviate poverty and unemployment among youth.
- Facilitate the development and strengthening of effective and efficient operations of youth networks and structures for good governance in Commonwealth member countries which are informed and fostered by the development and implementation of National Youth Policies (NYP).
- Develop and promote youth work education and training for the professionalization of youth work and its mainstreaming into the macro-development sector based on the body

of knowledge, skills, attitudes and core values and ethics.

- Assist governments in enabling young people to participate in society's economic, social and political governance.
- Disseminate and communicate world and Commonwealth-wide experience, results, lessons learned and best practices in youth development.
- Provide opportunities for young people to develop critical knowledge and skills in using information technology to create employment and share experiences by reducing the digital divide.

5.1.4 Capacity-Building Programmes Implemented by CYP

The support given to governments and other youth networks and structures by the CYP enables the member governments to refocus their resources on building their institutional capacity and develop comprehensive strategies that incorporate and mainstream CYP's programmes into the national youth development programmes of member countries. The support given to governments and other youth networks and structures by the CYP enables the member governments to refocus their resources on building their institutional capacity and develop comprehensive strategies that incorporate mainstream CYP programmes into member countries' national youth development programmes. CYP also conducts research on emerging youth issues and strategies and documents Commonwealth-wide case studies of best practices and lessons learned for dissemination to member countries for adaptation in national youth development programmes (Takawira & Simmons, 2004). The programme implemented by CYP are the following:

Youth Enterprise Development (YED) focused on creating decent employment and sustainable livelihood opportunities for young people to reduce the incidence of poverty and destitution in the member countries of the Commonwealth. The Commonwealth Youth Credit Initiative (CYCI) is an important component of the YED. It is a credit scheme which targets disadvantaged young people, specifically young women, addressing the circumstances and problems they face and ensuring that they establish successful and sustainable enterprises. The CYCI aims to contribute to national, regional and international efforts to reduce youth unemployment in Commonwealth countries.

The central objective of the CYCI is to provide a simple and sustainable model for micro-enterprise to equip young men and women with business management, information technology and entrepreneurial skills enabling them to establish viable micro-enterprises within their communities. It is a model of good practice, promoted and replicated in different countries with the support of international development and donor communities. The purpose of CYCI is to reach poor young people with no other access to credit (Takawira & Simmons, 2004; Ryan, 2006: Simmons, 2021, 2022).

Youth network and governance provide practical support for achieving Commonwealth values. This strategic programme area concentrates on enhancing the capacity of young people and youth networks to influence the governance process at all levels of society and enhance the development and implementation of National Youth Policies (NYP) (Joseph, 2006; Charles, 2006).

Youth participation allows CYP to enable young people to realize their potential as active citizens through their participation in, contribution to and engagement with development processes to strengthen democracy and human rights. Here governments and partner organizations are encouraged to include youth in their governance and management structures (Takawira & Simmons, 2004).

Youth work education and training focus on developing and promoting the profession of youth work through developing a body of knowledge, skills and attitudes, a code of ethics supported by the regulatory mechanism of occupational standards. In 1996 the CYP commissioned the development of a Diploma in Youth Work to be offered to Commonwealth member states through independent tertiary institutions. The diploma was accredited by over 30 tertiary education institutions in member countries and offered as supported distance education programmes to thousands of students across the Commonwealth. The diploma programmes utilize various innovative participatory methodologies and diverse training approaches, including workshops, work shadowing, mentoring and coaching, to deliver the training modules. It is now upgraded and offered by these institutions as full-fledged degree programmes (Maunders, 2006; Chanika Mataya et al., 2020).

It is important to note that as the Commonwealth and global youth development priorities shifted or funding allocations became more strategic relative to expenditure versus gains, several of the activities implemented by the CYP were scaled down and/or terminated, including the closure of the CYP regional youth centres.

5.1.5 Piloting Other CYP Capacity-Building Programmes

Commonwealth TEC

Young people are the pioneer of the ICT revolution both in terms of content and technology. However, although they are at the forefront of ICT, many young people do not have access to the Internet. Therefore the CYP has developed and piloted innovative programmes to enhance the capacity of young people in marginalized communities of the Commonwealth (Simmons, 2006). Over 3000 young entrepreneurs in remote rural communities in India have benefited from the CYP "TEC on Wheels", which is a mobile ICT training service. The government of Pakistan has requested the "CYP TEC on Wheels" to enhance the ICT skills of young people in rural and urban centres of Pakistan. The vehicle was delivered to the government of Pakistan in June 2005. Another innovative project implemented under this strategic programme is the CYP Technology Empowerment Centres (TEC) at CYP Africa and Asia regional centres. Under both initiatives, young people are using the new ICT skills to access markets and improve the operation of their businesses. A "TEC on Boat" initiative was piloted in the pacific region to provide access to technology in small remote island communities in the Pacific region.

The second phase of the new "CYP online" initiative was completed in 2004/2005. The project represents the most effective method of outreach to young people. Under this project, young people from Commonwealth countries who might not have any opportunities to travel outside their own countries can use the Internet to interact with young people and organizations from other regions of the Commonwealth. This model provides an opportunity for young people to participate in training programmes using the Internet and share experiences, lessons learned and best practices (Chanika Mataya et al., 2020).

Creating "Common Wealth"

Under the Youth for the Future initiative, endorsed by the Commonwealth Heads of Government Meeting (CHOGM) in Nigeria in 2002, a new model for utilizing enterprise development as a strategy for reducing unemployment and poverty among young people was instituted. The Creating Common Wealth (CCW) forum was implemented in November 2003 in collaboration with the government of Australia as a major pan-Commonwealth youth development event bringing 270 delegates from 45 member countries of the Commonwealth to participate in developing strategies and programmes to reduce unemployment. As a follow-up to the CCW initiative, short-term training projects in enterprise development were held in member countries of the Commonwealth.

5.1.6 Initiatives with Partners at the Global Level

CYP works with the Youth Employment Summit (YES) and the Microcredit Summit (MCS) to graduate over 100 million families worldwide out of poverty. In collaboration with YES, CYP supports (a) the development of YES country networks in sixty (60) countries which tap into young people's creativity, energy, idealism and innovation to assist member governments in achieving the MDGs, now the SDGs. This strategy also aims at moving young people into productive and long-term nation-building activities and away from non-productive pursuits; (b) the development of a stakeholders strategic approach, which defines the specific responsibility of each player/stakeholder, namely, young men and women, private sector, governments, donors and civil society institutions; (c) the creation of opportunities for CYP and the Commonwealth Secretariat to share experiences and models of its CYCI/Micro Credit Management System (MCMS), enterprise development and other poverty reduction strategies and programmes with young people, governments, donors and other stakeholders at the summit; and (d) the development and endorsement of an Alexandria Declaration on Youth Employment by seventy (70) governments including forty-five (45) ministers.

5.1.7 Cutting Issues

During the period under review, the CYP viewed globalization, gender, sports and environment and climate changes as critical cross-cutting issues that warrant serious consideration by the Commonwealth governments.

Globalization On the issues of globalization, instead of looking at youth in relation to the global economy, it is worthwhile to look at the effects of globalization on the individual and national levels, in terms of (a) migration flows within a country and abroad, (b) remittance flow from and to young people and changes in employment patterns, (c) access to microcredit and other forms of credits and (d) access to and usage of global media and Internet for education and capacity-building (Simmons, 2006).

Gender The issues of gender discrimination and equality are important considerations for measuring the performance of youth development. Young men and women are not merely a vulnerable group in society but are subjected to numerous development challenges due to discrimination. Therefore, it is important for CYP programmes to address progress in gender quality and should include the availability of a secured and adequate gender-responsive infrastructure, the number of young women in governance or authority roles and adequate access to training and productive resources for young women and men and reducing the incidence of the number of girls being exploited and being trafficked (Simmons, 2006).

Sports as a Tool for Development Sports and physical education play important roles in the development of young people and can contribute toward achieving SDGs. Sport and physical activities are critical considerations for measuring the performance of young people in youth development programmes, including focusing on sports in terms of (a) being used as a vehicle for re-engaging young people in education, (b) improving academic attainment in key skills such as literacy and numeracy, (c) increasing access to young women and girls in physical education, (d) the empowerment of young women through the development of leadership skills and confidence and (e) teaching young women valuable life skills through its inclusiveness.

Environment, Sustainable Development and Climate Change The environment has always been an important cross-cutting issue of CYP's strategic programme. Research has shown that climate change will continue to negatively impact all areas of the lives of young people and their communities, including their employment and livelihoods, health, education and social development (Simmons, 2018, 2021, 2022). In recent times CYP has developed a robust programme of action on climate change adaptation (CCA) which focuses on building the resilience of young people and their communities by enhancing their knowledge and skills and mobilizing them to take action to adapt to the impact of climate change (UNESCO, 2014; Simmons, 2022). The ability to adapt, that is, to maintain a responsive capacity, is grounded on three essential features, (a) the degree to which the system is prone to change while still retaining structure and function, (b) the degree to which it is capable of self-organization and (c) the capacity for learning (Nelson & Brown, 2007).

It is important to note that as development priorities change, the Commonwealth youth development priorities shift as funding is reallocated and/or becomes more strategic relative to expenditure versus gains. This shift in priorities resulted in several of the programmes implemented by the CYP being either scaled down and/or terminated, including the closure of the CYP regional centres.

5.1.8 Strengths of the CYP Capacity-Building Programme

The CYP has unique strengths, including providing technical guidance and global leadership on youth employment, governance and policy to member governments of the Commonwealth (Charles, 2006; Simmons, 2006; Joseph, 2006). The organization's vision focuses on harnessing the youth's capacity to amplify the governance of the emerging independent counties in the Commonwealth. The overall vision of the CYP is to "enable young people to realize their potential in the Commonwealth by transforming and improving the quality of their life, in partnership with other stakeholders, and in such a way, they are recognized as valued stakeholders and as part of their societies" (Takawira & Simmons, 2004, p. 85).

> The governments turned to CYP for guidance on youth development. The CYP implemented programmes which engaged the governments and Ministry of Youth Affairs officials on the role of youth in national development planning processes. CYP provides governments with the capacity-building and leadership in youth development, said Dr Henry Charles, acting director, Commonwealth Youth Affairs Division, Commonwealth Secretariat, UK.

This view was supported by Dwynette Eversley, programme officer of CYP regional centre, the Caribbean, who posits:

> CYP had the immediacy of reach and response to regional governments through Ministries of Youth by persons who are skilled and familiar with intricacies and development challenges confronting the regions. CYP had a key role in linking regional youth professionals, in sharing experiences and learn about good practices to replicate these different practices in the regions.

On the issues of CYP's unique strengths, Judith Webb, associate professor, UTT and former CYP development consultant, argues:

> CYP possesses unique skills and competencies to build strategic partnerships with government ministries, private sector agencies, multilateral organisations and other Comment Secretariat divisions to assist Youth Ministries and Youth Networks to mobilise needed resources and technical assistance to build the capacity to implement youth development projects and programmes.

The Commonwealth Youth Credit Initiative is a pioneering employment and sustainable livelihoods programme implemented by CYP. Florida Harrison, a young woman of the Chitekwere Group in Malawi, explained how she benefited from the CYCI:

> The loan from the CYCI has helped me to be independent and not to be a sex worker...I can buy my own soap, food, and clothes. I bought a parcel of land and am building my own house...these guys have nothing to offer me financially. I am even better off than most of them.

The key informants underscore the important role played by CYP in shaping and sharing youth development practices and policies across Commonwealth member countries in Asia, Africa, Europe, the Caribbean and the Pacific as the organization's unique strengths. CYP developed, piloted and rolled out numerous youth development programmes, including the Commonwealth Youth Credit Initiative (CYCI), the Commonwealth Diploma in Youth Work and Commonwealth Youth Micro Credit Management

System (MCMS). CYP implemented these programmes in member countries as successful youth development programmes. The CYCI was implemented in over 35 countries of the Commonwealth and was responsible for the creation of over 25,000 businesses during the first 5 years of implementing the project. Each of these businesses employed an average of 8–10 young people (Simmons, 2006; Ryan, 2006).

The CYP spearheaded work across the Commonwealth on recognizing youth work as a profession with its own values and competencies. The Diploma in Youth Work programme was implemented in over 30 tertiary institutions in 25 countries and was delivered to thousands of young professionals across the Commonwealth (Maunders, 2006). It was later transformed into degree programmes. CYP was identified by international development partners as one of the world's leading youth and capacity-building institutions, building a cadre of professional youth work development practitioners across the Commonwealth.

CYP responded to the request of member countries to enhance their capacity and leadership in youth development. It has employed a cadre of professional staff from the Commonwealth's respective regions and is familiar and knowledgeable with each region's specific youth development challenges. These staffs are committed to working with Ministries of Youth and youth networks to develop their youth programmes to resolve the development challenges confronting young people in the countries and the regions. A dedicated funding source supported CYP's work to implement its programmes.

The presence of the CYP at the regional levels in the Commonwealth was the key factor in influencing the uptake in youth policy development and another innovative youth programming by member governments of the Commonwealth. It also demonstrated the attempts by governments to improve the work and remits of *Youth Ministries and divisions* and mainstream youth development into national development plans and processes.

The CYP Youth Caucus (CYPYC) demonstrated an *excellent model for youth networking*, cooperation, good practice in democratic governance and building leadership skills and platform area competencies. The model has generated a *system of youth influencers and decision-makers* currently in positions of power which can reach out, cooperate and integrate other youth for regional and national development. The CYP effectively incorporated participatory strategies as a mechanism for mobilizing resources and building the capacity of youth community leaders, governmental ministries and departments at national levels. This strategy was also used by CYP to raise awareness and get youth to participate in activities at the community and national levels.

The work of the CYP was driven by research. During the period under review, CYP produces hundreds of strategic and thematic papers on diverse development issues. This strategy provides the mechanism for the organization to work with strategic partners to develop positive outcomes in policies and programmes in the Commonwealth.

CYP utilized various participatory innovative approaches to enhance youth development policies and programmes when youth development strategies were emerging as innovative strategies to transform the work of youth and their communities as vehicles to build young people's resilience. The CYCI was singled out as an essential initiative to build the resilience and independence of young women by preventing them from becoming sex workers and minimizing the potential of contracting HIV/AIDS. These women could provide for their families and the source of employment for other young people and the wider community. The strategy provides the mechanism for the organization to work with strategic partners to develop positive outcomes in policies and programmes in the Commonwealth.

5.1.9 Weaknesses of the Capacity-Building Initiatives

Like other development agencies, CYP faces critical weaknesses in its programming, management and operational structure. A key male informant who worked as a former senior officer of the organization states that the staffs of the CYP were not promoted based on merit. He said too many staffs were left behind to become frustrated and unproductive and concluded that the organization needs people who have the transformational leadership to drive it forward.

Dwynette Eversley, programme officer for CYP Caribbean programme manager, contended that more strategic directions should have been provided to regional staffs regarding programme development, piloting, implementation and evaluation:

> All components of the programmes should have been informed by tested good practice models and resources…in many ways, it was left up to the team lead's competencies to shape the development area.

The limited finance available and the limited number of staffs to implement the CYP programmes were critical issues raised by Judith Webb, associate professor, UTT and former CYP development consultant.

> Financial resources were extremely limited, and often programmes objectives were not fully achieved due to funding challenges and budget constraints… a limited number of trained and professional staff to do the work that was required affected outcomes.

Although CYP was involved in the creation of some of the most innovative youth development programmes globally, including the development of the Global Youth Development Index, the Diploma in Youth Work, the Commonwealth Youth Credit Initiative, the Micro Credit Management Systems and many others, the organization has its own share of institutional weaknesses. CYP is a component of the conservative, post-colonial entity, the Commonwealth Secretariat, which is deemed as the remnant of the old British Empire responsible for numerous atrocities such as colonialism and slavery in the countries now member of the Commonwealth. This reality leaves a vicious blemish on the name and brand of the organization. Although there were lots of efforts in the recent past to rid itself of its colonial heritage by changing the image of the organization, it is still viewed by most member governments as irrelevant and a colonial entity that "maintains the status quo" of the old British Empire. Barbados was deemed the most loyal country in the Caribbean to the old British Empire, which had rid itself of the queen as its head of state. This has triggered dialogues in several countries in the Caribbean, the Pacific and Africa to put the process in motion of becoming republics. During my 9-year employment at the Commonwealth Secretariat in the UK as deputy director of the CYP, I encountered several uncomfortable moments in some African, Asian and Pacific countries on behalf of the organization. I had to tolerate derogatory comments from senior governmental officials at me because I was representing an organization that has ties with a turbulent history which made the work environment very toxic and uncomfortable.

The CYP implemented considerable research in the field to assess the impact of its programme on young people in Commonwealth countries. The absence of a dedicated researcher and adequate funding to carry out such research to provide evidence-based findings continue to plague the quality of the organization's work. The organization also lacks robust and viable monitoring, evaluation and reporting (MER) systems to track these programmes' performance and provide the empirical evidence required to support the development and implementation of more viable youth development programmes in the field. The absence of empirical data negatively influenced the organization's internal working, mainly the development, management and operations of strategic programmes. It also negatively impacts how CYP supports partner agencies' work, especially youth departments, ministries and youth networks. The CYP collects data for preparing strategic and thematic papers submitted to the Commonwealth Youth Ministers Meeting (CYMM), the Committee of Management (COM) and other important Commonwealth governance meetings. Although these documents were of high quality, more attention should have been put to evaluating the outcomes of current programmes to extract the lessons learned and best practices to influence the formulation and development of the new CYP strategic programmes and policies.

The strategy of working in partnership with development partners to implement and roll out its work was a relatively new phenomenon for the organization. Although CYP worked best with Ministries of Youth, its work with international development agencies, such as UNDP and UNICEF, private sector agencies such as The Body Shop and the Commonwealth Bank, Australia, and civil society such as tertiary institutions, was generally ad hoc and haphazard. In today's development landscape, partnership and collaboration are critical for optimizing the results for the organization's end users. The core CYP resources, such as the *Youth Policy Handbook*, the *PAYE* and CYCI manual, should have been updated and made available for the next generation of users as valid good/best practices (www.yourcommonwealth.org).

Although CYP had a dedicated funding source, the financial resources to implement the programmes were minimal, and often programme objectives were not fully achieved due to funding challenges and budget constraints. The funding to CYP was a pledge by Ministries of Youth and senior officials of member governments. The funding was voluntary, and most governments were not compelled to pay their voluntary contribution to the Commonwealth Secretariat to support the operationalization of the CYP programme. The uncertainty of the funding of the CYP was the main factor which limited the outcomes of the organization's programmes. This situation was compounded by the limited number of professional staffs the agency could hire to implement the programmes at the pan-Commonwealth and regional levels.

The secretariat introduced a staff rotation policy programme with immediate effect, which states that staff at programme management level and above could only stay in post for a maximum of 9 years (three 3-year terms). Those senior staff close to 9 years had to leave the organization, and no effective mechanism was instituted to retain the organization's institutional memory. This led to the haemorrhaging of the institutional capacity of the organization. The organizations should have (a) hired the senior staff at least a month before the other staffs were leaving office to ensure effective handing over of files, sharing of knowledge and best practices on the workings of the divisions in the organization, (b) provision of moratorium of 1 year before effecting the policy to allow for effective transfer of knowledge on the workings of the organization and for HR to make the necessary arrangements to capture the institutional memory of the organization and (c) organization strategic workshop, Zoom meeting and others for new and older staffs to share information to capture the institutional memory of the organization.

5.1.10 Capacity-Building Challenges Encountered by CYP

CYP has its own set of development challenges. The decision-making structure for the CYP was complex and complicated and was not informed by the situation on the ground. They were made by the Board of Governors. The Board of Governors members were Commonwealth diplomats based at the high commissions of their respective countries in London, UK. These decision-makers were detached from the work of CYP and the Ministries of Youth in their respective countries and were making such decisions in a vacuum.

There were variances in opinions between the government officials in the capital and diplomats making decisions in London on the policies, programmes, management and finances of the CYP. The decisions were taken without feedback or follow-up with the Ministry of Youth Affairs officials in countries responsible for implementing policy and actions evolving from these meetings. The officials of the Ministry of Youth, when informed by CYP officials of the commitments to the CYP made by their diplomats in London, were startled since it was the first time they were informed of such commitments. They would complain about the lack of consultations between the diplomatic and the officials of the Ministry of Youth Affairs. This was the main reason for countries not paying the annual financial pledges made by diplomats at the COM meetings in London, UK.

Member governments are experiencing severe challenges mobilizing adequate financial resources to implement their development challenges; therefore, there are limited financial resources available for governments to implement mandates. Ministries of Youth and senior officials would deny making such a financial pledge at the Board of Governors and Commonwealth Youth Ministers Meeting (CYMM) upon returning to their country after such meetings, so CYP officials began electronically recording the entire meetings. Despite recording their voices, some senior officials would deny that they made such pledges even when they are sent the transcript of the meetings.

The CYP invested considerable time developing triannual strategic plans and programmes. However, rather than focusing on the strategic level, the CYP dedicated considerable time to tracking operational issues at regional and pan-Commonwealth levels. Hence, CYP did not spend sufficient time providing intellectual and technical analysis and guidance on the youth development programme implemented by member governments.

The contracting nature of the staff at the professional and supporting level was a continuous challenge for CYP. The contracts with staff at the regional centres were based on the salary of civil servants within the country. There were variances in the salaries of the staffs working at the different CYP regional centres. This was compounded by the difference in salaries of officials at CYP regional centres compared to the salaries of the staffs working at the Pan-Commonwealth Office in London, UK. This makes it extremely difficult for CYP to attract and maintain professional staff at the CYP regional centres. The wide difference in salaries of professional staff continues to negatively impact the organization's overall performance.

5.1.11 Strategies for Resolving the Capacity-Building Challenges

CYP realized that funding to implement the project was the organization's biggest development challenge due to the many governments not paying their financial pledges. CYP regional staff would meet with senior officials of the Ministries of Finance to encourage them to pay their financial pledges to the organization. The Commonwealth Secretariat would threaten the governments by taking away their voting rights at significant meetings and stopping implementing programmes in their respective countries if they did not pay the arrears for the pledges. During the period under review, these threats had little or no impact on the governments due to the few countries receiving from the Commonwealth Secretariat.

The CYP would collaborate with a range of partner agencies to mobilize £3 million in extra budgetary contributions (in kind) from member governments, private sector agencies, civil society (universities and NGOs) and international development agencies such as UNDP and UNICEF. The CYP collaborated with universities and colleges to develop the curriculum and implement the Diploma in Youth Work training and to put the occupational ethics and standard in place to develop youth work as a professional occupation. Similar work was done in all the strategic programme's areas, including Youth Enterprise Development. In youth entrepreneurship and employment, CYP piloted and tested CYCI and the MCMS in the four regions of the Commonwealth and then implemented the sustainable livelihoods programme in over 25 member countries.

The CYP added value to the Commonwealth Youth Ministers Meeting (CYMM) by organizing capacity development training for Ministry of Youth Affairs and senior officials of member governments in areas of youth policy development, strategic planning and programming, including youth employment and the CYCI, building the capacity of youth worker through education and training and standards of youth worker education. The CYP collaborated with UNISA to create a journal dedicated to youth and development. The *Youth and Development Journal* is dedicated to building youth and development capacity in the Commonwealth. The journal documented youth development research, sharing lessons learned and best practices on youth development among policymakers and practitioners. The journal is an important vehicle for building a relationship with universities. The Diploma in Youth Work practice was an exceptional capacity development initiative by CYP.

Elizabeth Namwala, now 29, is a mother living in Chipata Compound, a low-income housing estate in Lusaka. She received her first loan from the CYCI programme. She received a loan of K300, 000 from PFL to buy bread, maize and sugar. Before the loan, Elizabeth was selling 4–5 pans of buns each day. Now, she sells 50 pans of bread daily. She repaid her first and second loans before the stipulated time and is in the process of repaying the third loan of K800, 000. Elizabeth has expanded her business and employs five (5) young people. She has improved her home, and her general standard of living has improved immensely because of the additional income generated from the business. Elizabeth is a positive role model to other young men and women of the Chipata . *.Zambia*

5.1.12 The Capacity-Building Methodologies Utilized by CYP

CYP utilized numerous innovative methodologies to enhance the capacity of its staffs, policymakers and governmental officials, youth and other stakeholders. The emphasis was put on understanding adult education strategies and utilizing Paulo Freire's conscientization methodological approach for working with policymakers, youth development practitioners, youth and other key stakeholders to enhance their capacity in member countries of the Commonwealth. The agency utilizes other participatory education approaches, knowledge and skills to enhance the consciousness of young people. One of the shortcomings of the capacity-building programmes was that CYP did not spend enough resources (human and financial) on tracking and evaluating the effectiveness of the organization's participatory learning methodologies.

The organization effectively built regional and pan-Commonwealth networks and teams among Ministries of Youth and youth development practitioners along its strategic programmes. The networks and strategic teams were involved in implementing capacity-building training workshops on key content areas, project writing and resource mobilization, research on the gaps in the strategic programmes and strategies for sharing best practices and lessons learned in the critical area of youth development.

Besides implementing training workshops to enhance the technical competencies of policymakers and practitioners at regional and pan-Commonwealth levels, through Regional Advisory Board (RAB) meetings and the Commonwealth Youth Ministers Meeting (CYMM), capacity development training was implemented to enhance the skills and knowledge of the staffs on Youth Ministries on sustainable development issues. Professionals were brought in from outside the secretariat to train staffs on the critical development challenges affecting the organization and the development of strategies for resolving these challenges. The trainers utilize participatory and modular approaches. Such as group work, networking, coaching and mentoring to build capacity (Chanika Mataya et al., 2020). A staff development workbook was developed for training staffs of the CYP as a sustainable resource utilizing a modular approach to training. The initiative was implemented as part of 2-week dedicated strategic pan-Commonwealth workshops hosted by one of the CYP regional centres to (a) review the outcomes achieved from the programmes annually, (b) create the budgets and supporting policies for the next 3 years and (c) plan the strategic programmes for the next triennium. Overall, the train-the-trainers methodology invested by CYP enhances the staff's knowledge, skills and local capacity to replicate strategic planning training models at regional and national levels.

Although the training and capacity development methodologies utilized by CYP were effective. It did not recognize the full potential of the participants and their ability to build their capacity at the local level, the full benefits to be derived to the Commonwealth and the youth sector. Informants felt that the training provided by CYP should have gone toward assisting the staff in developing their own learning materials.

5.1.13 Lessons Learned

The lessons learned from the past CYP capacity development initiatives, which have relevance for the development of a new capacity-building strategic direction of the Commonwealth Youth Programme, are as follows:

- Involving young people at all levels of the capacity-building programme (from governance to operations) provides a sense of ownership and empowers them to perform critical partnership roles in national, regional and international development partnerships.
- CYP adds value to young people's development in the Commonwealth by enhancing the institutional capacity of the organizations and encouraging partner agencies, governments and other stakeholders to deliver better-quality programmes. This strategy enables government and development agencies to provide resources to support youth development programmes in member countries of the Commonwealth.
- CYP can improve its operations and processes and improve quality assurance mechanisms and processes. This improvement in the quality of CYP's capacity-building work resulted in attracting new partners (private sector, government, development agencies and institutions) to support its programmes in member states and at regional levels (partnership with The Body Shop in the Pacific region).
- Policies. CYP focused its work on creating opportunities for young people to participate in policy and decision-making through the National Youth Policy development programme. CYP has focused on creating opportunities for young people to participate in policymaking and decision-making through the development and implementation of the National Youth Policy development programme. It would continue its efforts to build robust monitoring and evaluation systems to provide adequate feedback to heads of governments and youth ministers on the performance of its programmes.
- Partners (private sector, governments and development agencies and institutions) will want to participate in youth programmes when they can see a clear demonstration of improvement of its operation and delivery mechanism, e.g. deepening the quality assurance mechanisms of programmes, creating opportunities for young people to participate in shaping and influencing policies and national decision-making processes, enhancing efficient management and strategic use of ICT to support communication, project management and building robust monitoring and evaluation mechanisms to provide adequate feedback on the performance of these programmes.
- The CYCI programme has demonstrated that young people are "bankable" (i.e. will repay their loans and participate in the credit programme) to boost their employment. If young people are encouraged as trusted partners and provided with training, mentorship support,

community outreach and credit, they can become effective and credible business leaders in the future.
- The CYP Youth Enterprise Development programme has the potential and capacity to enhance livelihoods and reduce poverty among people and local communities.

5.1.14 Best Practices

Numerous best practices are evolving from the CYP capacity-building programme, including the following:

- CYP has instituted the hybrid approach, utilizing the strengths of top-down and bottom-up approaches to implement capacity-building programmes in member countries. It also uses participatory methodologies to complement the delivery of its training and capacity-building programme.
- The CYCI and the MCMS have demonstrated over the year that microcredit is still a viable model to address development problems such as poverty and unemployment in Commonwealth member countries. The CYCI programmes have demonstrated that the youth microfinance projects receive repayment rates on loans ranging between 92 and 98%. Therefore, the model has shown that these entities are viable financial enterprises and that young people are "bankable".
- CYP sustained and enhanced its international reputation as a cutting-edge youth development agency; the target for extra-budgetary resources comfortably exceeded £3 million per annum, and its links with a range of development agencies and the private sector were strengthened.
- CYP's knowledge of building the capacity of member governments, youth and key stakeholders in youth microcredit, youth participation and National Youth Policy development has been successful and sought by a range of development agencies within and beyond the Commonwealth.
- The Commonwealth Youth Credit Initiative and Micro Credit Management System are in demand by multi- and bilateral agencies as mechanisms to reduce the impact of poverty among youth and meaningfully engage them in sustainable livelihood initiatives. These and other programmes of the CYP have maintained relevance by undergoing changes in content and the use of technology and other new delivery strategic channels.
- In the group solidarity model utilized by CYP, young people are requested to form groups of five to allow them to save and access loans to finance their businesses, which is compelling. It is viewed as the main factor which is responsible for the high level of repayment on loans.
- Gender, peace and respect for the law, the environment and climate change adaptation are cross-cutting issues for all programmes implemented by CYP.

5.1.15 Recommendations

The global financial recession is compounded by the COVID-19 pandemic and has devastated the economy and overall standard of living. Unemployment of young people is one of the most severe development challenges in the Commonwealth. It is compounded by the severity of the impact of climate change on young people and their communities retarding any efforts to build the resilience of young people and their communities to respond to this global calamity (IPCC, 2014a, b, c; Simmons, 2021, 2022; Nurse et al., 2014; UNDP, 2015). The situation is compounded by high recession rates in developed and developing countries. Governments are experiencing severe difficulties servicing their loans and mobilizing adequate financial resources to cover external pledges to develop international organizations such as the Commonwealth.

- CYP should kick start implementing a Commonwealth-wide consultation on youth employment, sustainable livelihoods and climate change. The pan-Commonwealth consultations will assess current strategies and best practices and explore new innovations and approaches to resolve the unemployment and underemployment challenges confronting young people in member countries. The CYMM would be an appropriate forum for exploring such youth development issues and may contribute to the enhancement of the financial pledges from Ministries of Youth and senior government officials.
- Although the CYCI/MCMS seems to have "lost its sting" as a cutting-edge CYP strategic programme, it is still deemed a viable strategic entity for tackling poverty problems confronting young people in the Commonwealth. There is an urgent need for a review of the CYCI/MCMS to make it more in tune with the current development needs of member countries. This should be followed by developing and implementing a pan-Commonwealth marketing and development plan to promote the CYCI. The emphasis of the CYP work should focus on building the capacity-building of youth and key stakeholders through the utilization of Information and Telecommunications Technology (ICT) as a tool for fostering poverty reduction, youth employment through entrepreneurship and sustainable livelihoods.
- The CYP should collaborate with UNICEF, UNFPA, WHO, universities and colleges to explore opportunities for the mobilization of financial resources to build the capacity of member governments, universities and youth networks to develop employment and sustainable livelihoods, climate change adaptation, youth policy, youth work training and governance initiatives in member countries. Resources should be invested to research best practices and develop strategies that can be adopted and replicated throughout the Commonwealth Youth Programme.
- CYP should create a youth development trust fund based on the social invest fund models to mobilize financial resources to support the development and implementation of youth development projects in the Commonwealth.
- The development, adoption and replication of monitoring and evaluation application software for all microcredit initiatives implemented by CYP. CYP strategic programmes (CYCI) should be a credible way forward. Therefore, it is critical for CYP to continue its international advocacy work on youth employment and sustainable livelihoods through the Microcredit Summit and the mainstreaming of youth development through the UN systems and the divisions within the Commonwealth Secretariat and across member countries in the Commonwealth.
- The development and application of robust monitoring, evaluation and reporting (MER) systems to enhance programme performance. Therefore, CYP should continue its advocacy work with governments on mainstreaming youth employment through entrepreneurship, youth policy development, environment and climate change adaptation and youth education and training through the programmes of the Commonwealth Secretariat and across member countries in the Commonwealth.

5.2 Case Study on Caribbean Youth Environment Network (CYEN): Providing a Voice for the Voiceless People of the Caribbean Region

This case study analyses the Caribbean Youth Environment Network (CYEN) capacity-building initiatives. It presents an assessment of the work of CYEN to implement capacity-building programmes to engage young people in environment and development initiatives across the Caribbean regions. The case study focuses on the strengths, weaknesses, gaps and challenges, strategies and methodological approaches utilized to build the capacity of young people and their organizations, the lessons learned and best practices of applying these strategic approaches and recommendations for enhancing capacity-building programmes in the organization in the future. The care study also focuses on the effectiveness of implementing bottom-up approaches to improving youth environmental programmes in English-, French-, Spanish- and Dutch-speaking countries in the Caribbean.

CYEN is a regional youth development network established by young people to work with young people approximately three decades ago. The headquarters of the organization is based in Bridgetown, Barbados. Initially, the organization evolved out of the work of the JEMS Progressive Community Organization in Enhams, St Vincent and the Grenadines, and the United Nations Environment Programme (UNEP). Three years after the UN declared "International Youth Year in 1985", a decision was made by UNEP to establish twelve (12) youth focal points across the globe to engage young people on environment and development issues. As president of JEMS, I was selected to participate in the global youth roundtable on the environment at UNEP geadquarters in Nairobi, Kenya. Simmons and the other young people met with Dr. Mustafa Tolba, executive director of UNEP and was appointed youth focal point for the Caribbean region (https://www.unep.org/). On return to the Caribbean in 1988, he set up the UNEP youth environment and development programme through JEMS and reached out to Calvin Howell, executive director of the Caribbean Conservation Association (CCA), for assistance in implementing a regional youth environment initiative. Reginald Burke, programme officer of CCA, was assigned by CCA to work with Simmons to explore the possibility of establishing a Caribbean-wide youth environment network to engage young people and coordinate their activities across the Caribbean (http://www.cfa--international.org/). After 4 years of consultative meetings across the region, the Caribbean Youth Environment Network was registered as a regional environmental organization that works with young people on behalf of young people on environmental issues across the Caribbean region.

Young people comprise approximately 45–50% of the population of the Caribbean (Carter, 2008). Young people are classified as the most vulnerable sector of the population to the impact of climate change and other development calamities (Simmons, 2021). Climate change severely impacts youth and their communities, enterprises and overall well-being, specifically in locations exposed to storms, floods, droughts and fires (CARIBSAVE, 2010, 2012; ILO, 2010, 2011; UNDP, 2015). Unemployment was estimated at 30–40% among the youth sector of the population but has risen because of the global recession, which caused many businesses to lay off staffs or limit the entry of persons into the job market. This global recession has exacerbated these challenges and has made it more difficult for the average young Caribbean citizen to prosper in the agriculture, fishery and tourism sectors.

The lack of opportunities for youth is a dismal calamity for Caribbean Small Island Developing States (SIDS) (Simmons, 2018, 2021). This problem is compounded by an increase in migration, especially among the most educated young people, resulting from being unable to envisage a future within their respective countries after not finding employment and/or adjusting to the social and political challenges facing their countries. Annually, thousands of young people would leave their countries in search of "greener pastures", mainly in Europe and/or North America. Such migration would lead to a brain drain, undermining national-level efforts to achieve Sustainable Development Goals (Mishra, 2007). It is estimated that the Caribbean countries have lost 10–40% of their labour force due to emigration to OECD member countries. Mishra (2007) contends that the migration rates are particularly striking for the high-skilled labour force. Several Caribbean countries have lost more than 70% of their labour force with more than 12 years of completed schooling. The Caribbean region has the world's highest emigration rates and is also the world's largest recipient of remittances as a percent of GDP, estimated at 13% in 2002. It is important to note that the economic loss due to migration to the OECD far outweighs the return in remittance and development aid to the Caribbean. Young people in the Caribbean also suffer from reduced formal education and training prospects within their respective countries (Carter, 2008; Simmons, 2018, 2021, 2022). Therefore, the migration of youth would continue to emerge as a major development problem in Caribbean SIDS in the future and would have serious consequences for achieving the Sustainable Development Goals (SDGs).

Who is classified as young people, and why are they essential to the Caribbean region achieving the SDGs? White (2012) states that understanding young people's lives requires looking at how youth are constructed, i.e. imagined and characterized as a meaningful social, economic and political group, and how youth is usually experienced by young people. He argues that policy discourse on youth development tends to view youth in a future-oriented way, that is, viewing youth (the person) as human capital and youth (the condition) as characteristically a period of transition (Simmons 2018, 2021). Thus, the definition of youth is an issue for many associations and countries in the Caribbean. Both the United Nations (UN, 1981) and the Commonwealth (Commonwealth Secretariat, 2007) have their age definition of youth. Youth is defined in the United Nations as a cohort of persons from 15 to 24 years of age, while the Commonwealth defines youth as the sector of persons aged 15 to 29 years (Simmons, 2006; Commonwealth, 2007).

The life cycle approach characterizes a young person (youth) as a phase of transition, i.e. when a person moves from a time of dependency (childhood) to independence (adulthood). According to the The World Bank annual report (2007), these transitions involve several common shifts, which present unique challenges for the young person and policymakers alike, including (a) moving from school to seek work and independent sources of income, (b) moving from parental home to new living arrangements and (c) forming close relationships outside of the immediate family, often resulting in marriage and children (Simmons, 2006, 2021; The World Bank annual report, 2007). The life cycle transitional approach is more appropriate for young people in the Caribbean because the approach acknowledges that young people's development varies at different stages and is influenced/shaped by economic, social and cultural circumstances. Therefore, a greater understanding of the specifics of each stage will be appropriate for CYEN and policymakers to design interventions for youth at each stage of the cycle.

There are two other fundamental development challenges facing young people in the Caribbean. These challenges involve (a) the lack of engagement in governance and development processes and (b) the lack of capacity development to empower young people and facilitate their participation in the above processes. Approximately 80% of young people do not engage in civic actions or participate in governance and national development. Young people are not being encouraged or supported to participate in decision-making processes in their respective countries even though their current generation and future generations will be impacted by today's decisions (Simmons, 2021). The problem is compounded by a perceived lack of urgency to deal with environmental and climate change issues and a lack of understanding of how current actions and decisions would contribute toward making the region unsustainable (CARIBSAVE, 2010; ILO, 2010; Nurse et al., 2014).

Capacity development within the organizations is the other critical development challenge faced by CYEN. Due to the diversity of new knowledge available from research on sustainable development and climate change, CYEN needs to be constantly focused on deciphering very technical information that may require assistance from experts with a background in these areas. Therefore, CYEN needs to ensure that its members are constantly provided with opportunities to represent the youth sector of the Caribbean at local, regional and international capacity-building events to bridge the information and skill gap. Therefore, all efforts on the part of the organization to equip members with good working knowledge of the region's environmental challenges and the ability to effectively communicate the Caribbean youth's position on these issues must be supported. While this may not be required for entry into the organization, the members' skills must be increased to adequately represent the organization and the youth sector at all levels. CYEN should continue to provide training to young people and encourage them to contribute to their own development by putting in the effort needed to research environmental and development issues to increase their knowledge and participate in decision-making at community, national, regional and global levels (https://cyen.org/).

5.2.1 The Mission and Vision of CYEN

The Mission of CYEN

The Caribbean Youth Environment Network is a non-profit organization dedicated to improving the quality of life of Caribbean young people by facilitating their personal development and promoting their full involvement in all matters pertaining to the environment and sustainable development. The organization promotes education and training, Caribbean integration and community empowerment as tools to develop an ethic among young people that assists in conserving and protecting the natural resources within the wider Caribbean (https://cyen.org/).

The Vision of CYEN

CYEN will be the leading youth environment and development organization in the Caribbean. It will be responsible for creating an appropriate space for young people to participate in the design, development, implementation and promotion of new, progressive, equitable, innovative and sustainable environmental, economic and social development policies and programmes across the wider Caribbean (https://cyen.org/).

5.2.2 Programmes Implemented by CYEN

CYEN's work is based on three development pillars. These pillars include (a) engagement and participation, (b) awareness and education and (c) advocacy (https://cyen.org/).

The organization recognizes that the key driver for change in the Caribbean is not through education but through the culture of young people; the organization can create the necessary changes to ensure the achievement of Sustainable Development Goals (SDGs).

> Education as a notion for creating change is based on a western perception of change. The Caribbean culture is viewed by CYEN as the most dominant driver for achieving change among young people. The people in the Caribbean got caught up in the western notion of change due to the exposure of the people in the Caribbean to the North American and European media. Reginald "Reggie" Burke, executive coordinator of CYEN.

The specific programmes implemented by CYEN evolving from the three development pillars emphasized above include the following:

(a) Climate change, sustainable land management and biodiversity.
(b) Integrated water resource management.
(c) Coastal and marine water resources management.

The Caribbean is a region through which the Caribbean Community for Climate Change Centre (CCCCC) adheres to the Paris mandate theme of "Moving beyond 1.5 C to stay alive". The focus is on building the people's resilience in the Caribbean to adapt to the impact of climate change. CYEN adheres to the mandate of the CCCCC and is working with young people in the Caribbean to promote and advocate for resources to foster the achievement of the SDGs (https://cyen.org/).

5.2.3 Strengths of the Caribbean Youth Environment Network

CYEN has unique strengths in working with young people in 15 Caribbean countries on environmental and sustainable development issues. It is a youth development organization created by youth for youth across the Caribbean.

> The organisation implements a regional environmental programme to attract youth across the Caribbean. This is a major strength of CYEN as it presented a cohesive front when engaging in a regional and international environment and development activities, posited Renee Boyce-Drakes, former chairman of CYEN.

Yoland London, former National Youth Coordinator (St Vincent and the Grenadines) and CYEN executive member, posited, "We were able to identify the environmental challenges facing the countries in the Caribbean... and provides training and support the mobilisation of young people to take actions in their respective communities to resolve these challenges".

Reginald Reggie Burke, executive coordinator, CYEN, explained the benefits of working with youth to enhance their capacity to become part of their country's national delegation.

> Getting youth representatives on national delegations of member governments in the Caribbean to attend regional and global meetings/conferences organised by UN, UNFCCC, the Commonwealth Secretariat, and other international development institutions was an accomplishment for CYEN. To support the realisation of this strategy, CYEN is working with Governments in the Caribbean through the CCCCC to ensure that policies are put in place to ensure youth are represented on national delegations. CYEN office in Barbados is working with the National Coordinators to get young people to engage with their respective government to be members of their government's delegation to these high-level regional and global meetings.

Through the CYEN capacity-building programme, the organization provides excellent exposure for young people to build comradery and friendship and share their experiences, knowledge and skills with other young people on the environment and critical development challenges in the Caribbean. Young people in the Caribbean would not have such opportunities if not for the organizations working on building their resilience to adapt to the impact of climate change and other development calamities. The organization provides possibilities for young people to build and share best practices in their actions to tackle environmental and climate change issues at the community and regional levels. It also offers the opportunity to share best practices in youth development and climate change adaptation with the rest of the world. Young people could utilize different tools, methodological approaches and strategies to assist them in enhancing their own life chances and contributing to communities and national development.

The organization provides avenues for experienced members to mentor younger members by sharing their knowledge, skills and experiences to resolve environmental issues, mainly climate change and other development issues, thus guiding young people to make decisions on critical issues regarding the organization's work. Often, young people are left out of decision-making processes by governments and other institutions in the Caribbean. CYEN provides opportunities for young people's voices to be heard on these critical issues of national significance. The experience in the Caribbean shows that, quite often, young people are taken for granted by society. "We want our voices to be heard because we want to be 'part and parcel' of the process of making a difference...yes, we want our voices to be heard", said a young female member of the CYEN who benefited from the capacity-building programmes implemented by the organization.

CYEN created opportunities for young people to hear their "voices heard" at several SDG international events. Young people participated in UNFCCC COPs, Commonwealth Youth Forum (CYF) held during the Heads of Government Meeting (CHOGM) and other global meetings/conferences on land management and desertification, biodiversity and others. "I participated in the UNFCCC CoP meeting in Copenhagen and was chosen to present the youth statement evolving from the youth forum to the Caribbean leaders attending the Climate change summit on how we, as young people going to tackle the climate change challenges in the Caribbean", said another young lady as she spoke of the benefits she received from the organization. Over the past decades, CYEN trained a cadre of young people who are capable of effectively representing their countries and region at high-level meetings. These young people were provided with knowledge, skills and competencies on climate change adaptation and sustainable development issues; negotiation and conflict mediation; monitoring, evaluation and reporting (EMR); communication strategies; strategic planning; and strategies for enhancing presentations at meetings.

CYEN was effective in incorporating students into the programme. A young female spoke of the benefits of participating in the Youth Environment and Service (YES) clubs project in St Vincent and the Grenadines. The YES initiatives were implemented in over 40 primary and secondary schools in St Vincent and the Grenadines. The purpose of the YES was to engage children and youth in environmental programmes in schools and communities. The YES initiative was implemented in partnership with JEMS and involved other partners such as the Ministry of Education, principals and head teachers of primary and secondary schools, business houses such as DIGICEL and FLOW and community leaders and their organizations. The YES initiative implemented numerous community clean-ups, tree planting programmes, training workshops on the environment, environmental camps and other environmental actions to raise awareness of the issues relating to the degradation of the environment and to build the resilience of their communities. The YES concept was transferred to the other Caribbean and African countries, the CYEN and UNEP. The YES initiative was testimony that people were motivated at community and national levels to be part of the solution to address these global challenges through their actions within their community. It is an excellent example of "thinking globally but acting locally".

Its members' adherence, commitment and dedication to the values and principles to protect and conserve the environment is a major strength of the organization. CYEN has developed a holistic, integrated approach to resolving the sustainable development issues facing people in the Caribbean. Over the past two decades, CYEN has acted as a conduit for international development agencies such as the UN, World Bank and others to convey critical information on strategies to engage the youth sector of the Caribbean in sustainable development activities. The structure of the organization is designed to convey information to the youth sector within 2 min to all the countries in the Caribbean. The unique dynamic model enhances the participation and empowerment of young people in the governance and decision-making processes at the community and national levels. The UN is currently working with CYEN to explore the possibilities of setting up similar entities in Africa and the Pacific regions.

There are numerous deficiencies in tertiary education officered to students across the Caribbean. The degree, diploma and certificate courses are highly technical in terms of the skills to be acquired by young people; however, they are very limited in terms of the rate of transfer of the "softer skills" required for employment. There are numerous young people who acquired degrees from accredited tertiary institutions in the Caribbean who are finding it difficult to write business letters to complain to their parliamentarians, write applications for a job, prepare a newspaper article on a project, manage their time, develop strategic and work plans, prepare project proposals and write M&E report. The organization trained members in these skills through an internship, coaching, apprenticeship, mentorship and workshop. CYEN is viewed by many governments and development organizations as a "finishing school" for young people desirous of working in environment and development. The organization provides opportunities for young people to sharpen and fine-tune these critical skills to make them employable.

5.2.4 Weaknesses Identified by the Organization

The two most critical challenges confronting civil society organizations in the Caribbean are the lack of adequate finance and the low capacity to run the organization's programmes. These factors affect the organization's ability to grow and expand its programmes.

> The most critical weakness of CYEN is finance. This is compounded by environmental and climate change issues that are not on the 'front burner' of decision-makers, the cooperate sector and most development institutions in the Caribbean. Despite

> all the challenges regarding the impact of climate change on small island communities, it is still not appealing to potential sponsors and therefore, when you are working on environment and climate change issues, you must work much harder to sell your projects to proposed sponsors. Yoland London, National Youth Coordinator (St Vincent and the Grenadines) and CYEN executive member.

Negative attitudes of people who evolved out of the colonial past in the Caribbean were viewed as a significant factor affecting the performance of the organization.

> People in the Caribbean have a flawed mindset that impacts everything we do regionally and nationally. There is a tendency for people to push their national issues over the interest of the regional Caribbean-wide issues. There is a divide-and-rule mentality that evolves out of the region's colonial past. This notion is compounded by colourism, i.e., a tendency for us to promote and or favour persons of a 'lighter complexion' irrespective of whether the person possesses the competency and attitude to get the work done. Reginald "Reggie" Burke, executive coordinator, CYEN.

Finance is needed to support the work of the organization. This was emphasized as a very critical issue facing the organization. A male beneficiary of the organization said, "there are so many projects for the organisation to implement in communities across the Caribbean, but it does not have the financial resources to implement such projects". The lack of funding to support these projects frustrate the leadership of the organization. The enthusiasm was very high at the community and national levels to implement climate change adaptation projects, but there was no money to support the implementation of climate change adaptation projects in the community. Generally, although there are lots of finances available at the international level to fund climate change and environmental projects, the funding provided by development agencies such as the UN agencies, World Bank and Global Environment Facility (GEF) is not provided to NGOs to support the administrative cost incurred by the organization to implement such projects. The donors do not give funding to pay for the coordination, communication, rent, management and other administrative costs to ensure smooth project implementation. This does not make sense because poor youth NGOs cannot effectively implement projects if they do not have the money to pay to employ professional staff to manage these projects in the field, communicate with all stakeholders and prepare the necessary reports for submission to the donor agencies. There are concerns raised by informants that organizations with leaders with a lighter complexion in their leadership would most likely receive funding for their projects from the donor. This issue requires further research to determine whether there is any basis for such an accusation.

The Caribbean has very stringent financial regulations due to efforts by governments and donors to hinder incidences of corruption and terrorism. The governments in the Caribbean are signatories to global anti-corruptions and terrorism legalizations to access the global financial systems. Therefore, civil society organizations in the Caribbean must comply with these laws if they are to benefit from the funding provided through donor agencies. The young people do not have the confidence to deal with such financial compliance due to fear and ignorance of these financial obligations and the working of the financial systems. Young people in the Caribbean lack information on such critical development issues and do not apply for funding to support the implementation of climate change adaptation projects on the ground due to fear that they may get in trouble if they violate these regulations. This situation is compounded by the fact that the leaders in the Caribbean do not have the confidence in the youth to manage public funds transparently and do not create the necessary mechanism for their organization to access such donor funds like other sectors of the society.

The non-inclusion of young people in the decision-making process in the Caribbean is viewed as a major issue facing communities and countries. Decisions are implemented by government and local authorities through a top-down approach. Government officials would decide how to implement a project in their offices and expect young people and other marginalized sectors of society to participate in implementing these projects with little or no consultation with the youth sector. The top-down approaches, as practised by government agencies, stifle the participation of young people in the decision-making process (Matland, 1995; OECD, 2013). When decisions are made without the input of young people in the decision-making process, it suppresses the growth and development of the young people and the entire country. "These decisions are being made without our input, and they result in causing lots of problems. Therefore, it is necessary for us as young people to participate in finding solutions to the challenges affecting the community and the country", stated a young male. The work of CYEN over the past decades shows that young people have what it takes to be part of the decision-making process, i.e. they have the knowledge, skills and capacity to participate efficiently in making decisions and contribute to the national development process.

Most young people see CYEN as a social club like the Lions Club, Jaycees, etc., as opposed to a civil society organization committed and dedicated to building the resilience of young people and their communities to adapt to the impact of climate change and the achievement of the SDGs. Most young people are not taking their commitment to the organization seriously; they will quit the group as soon as they experience difficulties with the other organization members. This situation is compounded by a culture in the Caribbean where political and civil society leaderships do not take young people seriously, even when the young person possesses the academic qualification and the necessary competencies to be treated with respect.

5.2.5 Challenges Identified by the Organization and Communities

Numerous challenges identified by CYEN limit and/or retard the organization's operations. These challenges may have implications for the overall and effective functioning of the organization.

> Since climate change and environmental protection are not as attractive as sports development etc., getting and maintaining membership of numbers of the organisation was a major challenge, contends Renee Boyce-Drakes, former chairman of CYEN.

During the early stages of the organization, members could not afford the cost of technology and the Internet to support the organization's smooth operation.

> The access to technology through the internet was a major challenge in the earlier time of CYEN. It was too expensive for ordinary people to purchase a computer and pay for the internet continuously. The telephone costs were too expensive, so it was difficult to basically keep in touch with colleagues and their organisations for any sustained period, posits Yoland London, National Youth Coordinator (St Vincent and the Grenadines) and CYEN executive member.

A large percentage of young men in the Caribbean are disengaged from society.

> There is a critical challenge in the Caribbean regarding engaging young men in the environment and climate change adaptation actions. There is an 80/20 ratio in terms of women's membership in the CYEN. Young women are fully engaged in education, join organisations, and are involved in other productive activities. While there is a tendency for young men to be engaged in issues that do not focus on their personal and/or economic development. Some young men waste lots of time hanging out on the 'block' and are involved in crime, drugs trafficking and other illicit practices, stated Reginald Reggie Burke, executive coordinator, CYEN.

The lack of finance to support the development, implementation and management of climate change adaptation projects at community, national and regional levels is the most crucial challenge identified by key informants in the Caribbean. This is compounded by the absence of sponsorship to support young people's participation in regional and global meetings on environmental and climate change issues.

It was a challenge getting young men to join CYEN because environmental and climate change issues were not deemed "sexy" and not on the priority list of young people. Young men have more interest in music and technology and are also distracted by the incidence of crime, drugs and illicit sex (Simmons, 2006; Charles, 2006). Young people would exhibit interest in the environment and climate change issues only when there are threats of hurricanes and violent storms in the Caribbean. Then they will lose interest when the imminent threats subside. A former male member of CYEN contends, "despite such challenges, we never give up working with young people…trying to persuade them to be part of the organisation... once we got them to become members of CYEN, we were able to build and sustain their interest in the environment and climate change issues by introducing them to community 'five a side' football, netball, 'tug-ah-war'[sic.], and other sporting activities".

Due to the excellent role CYEN is playing in engaging young people in resolving the development issues in the Caribbean, governments should mobilize funding and other resources to fund the activities and projects of young people in their communities. The governments committed to international organizations such as the UN and the Commonwealth Secretariat to provide funding for young people to participate as members of their national delegations to meetings/ conferences. There is no doubt that the participation of young people in these meetings would contribute to building the capacity of young people to contribute to the national development processes.

An assessment of the work of CYEN shows that young people are desirous of participating in environmental and sustainable development projects but lack the finances and technical know-how to achieve such a goal. The governments in the Caribbean should forge a partnership with CYEN to enhance the technical and management capacity of the organization by providing mentorship, coaching, apprenticeship and other capacity-building support to the CYEN national chapters within their respective countries.

The informants believe that there is still a high level of mistrust of young people by governments and donor agencies regarding the use of finances to implement projects. This may be due to government and donor officials' lack of confidence and trust in young people regarding properly utilizing project funds. This may be due to practices in the Caribbean regarding young people's poor management of project funds.

The issue of culture is critical and relates to the behaviour and practices of people about protecting the environment. There is a practice among Caribbean people to dump garbage anywhere. Informants believe that governments and private sector agencies are the main perpetrators of this problem.

5.2.6 Approaches to Resolve the Challenges Encountered by the Organization

CYEN has implemented numerous participatory innovative strategies to resolve the development challenges identified within the organization and across the Caribbean region. Regarding the disengagement of young males in the Caribbean, Renee Boyce-Drakes, former chairman of CYEN, states:

> Membership drives were carried out continually targeting out-of-school youth, secondary and tertiary students.

Concerning the issues of funding of CYEN, the organization implemented innovative approaches to build the confidence of donors and governments to provide funding to support the organization's work. Reginald "Reggie" Burke, executive coordinator, CYEN, shared that:

> Donors do not like to give young people money. Having older professionals working within the organisation adds weight and convinces the donors who are giving the money to support the organisation's work across the Caribbean. This is compounded by the established track record and history of CYEN to implement innovative environmental projects among the youth sector.

Yoland London, National Youth Coordinator (St Vincent and the Grenadines) and CYEN executive member, presented strategies to address the funding challenges:

> To address the issue regarding the lack of funding to support our projects, we did fundraising activities in communities and at a national level. The

people see what we are doing, and this attracts them to come on board to support our community climate change and disaster risk management programmes. The community and private sector companies step in to support our clean-up and climate change adaptation activities. We have businesses such as Carib Cable, DIGICEL and FLOW supporting our projects over the year.

CYEN instituted various innovative strategies to address the organization's capacity-building challenges. The first strategy to address the financial challenges was through donations from members of the organization who contributed their own funds, food and other in-kind contributions to support the community cleaning, tree planting and climate change adaptation projects. Other affiliates, such as JEMS, collaborated with CYEN to develop and implement training of trainers programmes for leaders in project writing, resources and community mobilization strategies, climate change adaptation and mitigation, climate risk management and other technical and practical approaches to build the capacity of the organization. Such skills allow the organization to prepare sound project proposals to submit to funding agencies such as UNDP GEF Small Grants for funding. Donors in the Caribbean do not like to give funding to young people because they don't have the confidence that these young people will utilize the funding effectively. Therefore, having older mature professionals working with the organization enhances the confidence of the donors and private sector agencies to work with the organization.

CYEN was consistent in its approach to addressing environmental and climate change issues. In demonstrating such a commitment to resolving the environmental challenges, one young lady posits, "Our commitment to developing climate change adaptation projects and programmes was second to none. We were not willing to give up even when funders were telling us that they were not going to fund our programmes...we continue to develop strategies to repackage and promote our environment and climate change projects to people at a community level, governments and businesses, and it paid off in the end – big time. CYEN also took a business approach to brand its environment and climate change adaption projects and market them to the countries as viable products".

The culture in the Caribbean has changed drastically during the period under review. The community spirit that was part of the youth sector in the 1980s–2000s is now missing in the sector. Everybody wants to be their "own boss", e.g. during that period hosting regional youth environment activities was very cheap in that you could have put up to four young persons in the same room to sleep, nowadays everyone wants their own room. This is making it more expensive to host and organize regional and national activities. Young people in the Caribbean are more individualist, thus the reflection of the political culture in the Caribbean region. "Young people in the Caribbean talk a lot, and their talks are supported with little action", posited a female member of CYEN. However, due to the urgency of the climate challenge in the Caribbean region, there needs to be less talk and more action on the part of young people, community leaders and policymakers. On the flip side, the culture relates to the arts, design and music; CYEN is capitalizing on this as a medium to send messages regarding the negative impact of climate change and the need for urgent actions. The organization has created videos and jingles utilizing the dominant youth culture in the Caribbean, the music and the fashion, to raise awareness on the issues relating to the environment.

The UN and other international development organizations operating in the Caribbean would select young people to attend high-level international meetings due to how well they perform in an essay writing competition. Others are selected for these international meetings based on their relationship with a senior government official or the organization's staff member. The young people are not members of any youth organization or network, so there is no organization for them to report to and/or share the skills and knowledge gained at these meetings. CYEN is working with the UN agencies and governments to persuade them to select young people who are members of an organization to participate in regional and international meetings/conferences. This will facilitate the transfer of knowledge and skills to other young people in their organization and beyond.

5.2.7 The Capacity Development Problems Faced by CYEN

Youth environment organizations in the Caribbean cannot develop and implement environmental and sustainable development projects and programmes.

> Opportunities for training members presented a challenge in the organisation when trying to ensure all the youth leaders were able to access training. Renee Boyce-Drakes, former chairman of CYEN.

The organization emphasizes volunteerism rather than mobilizing resources to hire staff to implement projects and programmes to build the organization's capacity.

> The organisation's work was 100% voluntary, and it did not have any full-time staff to run the projects and programmes. The lack of full-time staffs and experts to develop and implement new projects was a major challenge facing CYEN. There was too much focus on voluntarism rather than on mobilising financial resources to pay staff and support the delivery of projects. Members were preoccupied with attending to their full-time jobs and did not have the time to track the organisation's operations. This causes lots of projects and programmes to run a drift. Several members of CYEN were unemployed and did not have the money to pay for transportation to attend meetings, argues Yoland London, National Youth Coordinator (St Vincent and the Grenadines) and CYEN executive member.

The complexity of the language problem in the Caribbean continues to negatively impact the organization's performance. There are four distinct languages spoken among the people in the Caribbean.

> CYEN still needs to find ways to break down the language barrier which prevents smooth communications among the members of the organisation across the region. The language predominantly spoken among members are English. Therefore, there are difficulties in communicating with Haiti, which is French-speaking and the Dominican Republic, which is Spanish. Most of the people and islands in CYEN are English speaking, and although there are more efforts from young people to bridge the language barrier, the communication problem still exists, commented Reginald "Reggie" Burke, executive coordinator, CYEN.

Communication through appropriate and acceptable language and staff shortages are the organization's main capacity development challenges. CYEN has expended many efforts to train young people in technical areas relating to climate change adaptation and sustainable development to boost the capacity of members and their networks.

It has been a challenge for CYEN to develop and implement robust monitoring, evaluation and report system. In reflection, the organization did not spend enough time documenting the implementation processes of its projects and activities. A considerable amount of time was spent on executing the project activities; however, not enough time was spent documenting the processes leading up to achieving the project outcomes. A former executive member of CYEN felt that although these projects created the necessary impact on young people and their communities, enough time was not spent documenting the projects' processes and outcomes. She concludes that the members of CYEN were driven by their desire to complete the project activities within the stipulated time frame rather than ensure the processes were documented.

Over the past decade, CYEN has trained many volunteers to develop their capacity through online courses and physical workshops. Still, the limited financial resources available to implement this training thwarts the organization's ability to provide opportunities for graduates to utilize their skills to implement a project at the community and national levels. Regarding the immediacy of the climate change problems in Caribbean SIDS, there is a need for civil society actors such as CYEN to "step to the plate" to resolve the environmental and climate change challenges facing the region. Civil society organizations are not as active as in the 1990s and 2000s. It is believed that the inaction in the civil society sector in a period when its work is more needed is due to the sector lacking leadership and vision to take the climate change adaptation endeavours forward.

There are political factors that affect individual countries and regions. These external factors further act as mechanisms to divide the young people and affect their participation in CYEN and other national and regional organizations. Besides the political issues marginalizing young people, their organization is experiencing difficulties mobilizing young men to participate in the programmes of CYEN. Young men across the Caribbean region are not engaged in the community and national development programmes. This is due to the large percentage of young men who are not engaged in education and development programmes (Charles, 2006; Carter, 2008). It is also evident that most young men who are qualified in environment and development would eventually leave the organization as soon as they are employed, further limiting the transfer of knowledge and skills from these young persons to other youth, particularly the organization.

The small population of most Caribbean states and the limited opportunities available for qualified young people in the islands further create barriers for young people to stay in these islands. Most young people will leave their island as soon as employment opportunities are available. For example, many young men and women from CYEN Grenada travel to the UK in search of employment in the UK Army and Navy. The migration of youth leads to brain drain and haemorrhaging of the limited human resources in the country. It also led to the non-function of the CYEN in Grenada for a considerable number of years (UNDP, 2015; Simmons, 2021).

The youth in CYEN are also very myopic and nationalist regarding how they respond to issues that affect the organization. Most young people would recommend their fellow nationals to represent the organization at high-level meetings and conferences rather than another person from another island in the Caribbean to ensure that the best qualified and skilled persons represent the organization. Young people are less committed to serving their country and the organization than youth two decades ago. There is a belief by the current cadre of young people that the coordination of CYEN is an easy activity. However, the current youth would rather resign when they experience difficulties rather than continue with the organization. Young people are not engaged in the management and implementation of sustainable development projects and/or organizations in their country, so they lack the experience and toughness required to manage an organization. CYEN is organizing training, coaching, apprentice, internship and mentorship activities to groom young people to manage such an organization.

5.2.8 Strategies for Resolving the Human Capacity Development Problems

CYEN devises innovative strategies for addressing the capacity-building challenges the organization is encountering. Such strategies include working in partnership with other development organizations in the Caribbean region.

> We reached out to funding agencies to assist with arranging opportunities for training youth leaders. They help us to set up online training events to promote the work of the organisation, states Renee Boyce-Drakes, former chairman of CYEN.

Training workshops were highlighted as a key strategy implemented by the organization to build the capacity of its members:

> We were very creative in coming up with strategies to address the organisation's human capacity-building problems. We organised training workshops for a) the persons who attended meetings and other training in the country and internationally to share the skills and knowledge acquired with members of CYEN and the community, and b) skilled officials from government and private sector agencies were invited to share new knowledge and skills at these workshops and c) skilled young people from CYEN regional networks who implemented innovative projects in their countries shared the outcomes of their work with colleagues, explains Yoland London, National Youth Coordinator (St Vincent and the Grenadines) and CYEN executive member.

The CYEN utilizes music and other cultural art forms as a vehicle to aid the transfer of new knowledge and skills to young people:

> We at CYEN work with young people to make things culturally relevant. For example, when spreading the message on environment and climate change adaptation, we prepared a 20-second jingle that is based on the culture of the people of the Caribbean, i.e., linking into the music, arts, and design of the Caribbean. I remembered how effective it was when JEMS used drumming music as part of its Moonlight Cultural Wake Model in St Vincent to raise awareness and enhance the participation of people in the environment and climate change campaigns. Reginald "Reggie" Burke, executive coordinator, CYEN.

Young people are not sitting on the periphery of the movement to achieve sustainable development in the Caribbean. They are now at the centre of the process of finding solutions to resolve the environmental challenges confronting the countries of the Caribbean. CYEN advocates for implementing a holistic, integrated participatory approach that this approach incorporates people as an integral part of managing the natural resources of island nations. It calls for all stakeholders to work together to manage the environmental challenges. The CYEN bottom-up management model views the end user as part of the same management mechanism and not as a consumer and/or destroyer of these islands' natural resource base. The concept is new to most young people and policymakers in the Caribbean, so CYEN is putting more emphasis on training young people to understand the model so they can be advocates for the model at the community and national levels. It also trains youth leaders and advocates to develop softer skills such as writing job applications, newspaper articles, negotiations and conflict resolution. These skills critical to enhancing young people's careers are not being taught in universities and colleges as part of the career guidance package (Simmons, 2022).

The organizations also concentrate their limited resources on the peer-to-peer transfer of knowledge and skills among young people. This strategy concerns young people from CYEN country offices working with other young people from the CYEN regional office to facilitate the training of young people in new skills and knowledge and strategies for resolving environmental and climate change issues. The strategy is based on the organizational capacity-building philosophy of "each one teaches one". Another important innovation is that CYEN ensures there are representatives of the organization in all training workshops on the environment and sustainable development issues at national and regional levels. Over the years, the organization adopted a "having its ears on the ground" approach in search of capacity-building opportunities for young people. This strategy has been very beneficial to the members of the organization. CYEN would receive invitations from partner organizations to attend regional and international meetings and conferences. Due to the high standards used to select delegates to attend these workshops/conferences, the host organizations would recommend these young people to attend similar international activities in the future.

5.2.9 Capacity-Building Methodologies Are Used by CYEN to Train Young People

The organization utilizes innovative methodologies to build the capacity of young people, its members and their communities to tackle environmental and climate change challenges in the Caribbean.

> CYEN organises Online training events in collaboration with regional and international agencies. Renee Boyce-Drakes, former chairman of CYEN.

The organization organized workshops and conferences as its premiere methods to enhance the capacity of young people.

> The chief among the methodologies used by CYEN are workshops and conferences on environment and climate change issues to enhance the skill, knowledge and competencies of the members and community leaders on the extent of the impact of climate change on communities. We also train young people on the strategies that can be implemented to enhance the resilience of young people in small island states to adapt to climate change. Workshops were also organised on project proposal writing to mobilise financial resources to support the work of the organisation. Reginald "Reggie" Burke, executive coordinator, CYEN.

Mentorship and coaching were also utilized to enhance the capacity of the members of VYEN:

> Mentorship support, coaching and peer-to-peer approaches are some of the methods used by CYEN to build the capacity of young people. I like the peer-to-peer approach because it brings young people to work with others within the same island and across the islands of the Caribbean to support and facilitate the transfer of knowledge and skills. Yoland London, National Youth Coordinator (St Vincent and the Grenadines) and CYEN executive member.

CYEN encourages its members to use online courses run by the UN and other partner agencies to enhance the skills and knowledge of its members on environmental and climate change issues. It also runs online courses for its members on similar topics (Bloomfield et al., 2017). It invested considerable time and resources to organize seminars and workshops on climate change, land management and other critical issues to train its members and community leaders. Although workshops and conferences are more expensive to implement than online courses, they have unique advantages in that you are sure that persons are there in person participating in the training exercise. There are more interactions among young people when they attend workshops and conferences, which leads to the building of partnership and comradery among young people (Chanika Mataya et al., 2020).

When utilized during capacity development training, the participatory development approach ensures a smooth transfer of knowledge and skills to participants (Chambers, 2002). The young people attributed this smooth knowledge and skill transfer to their active participation in the training activities. The bottom-up participatory methodologies used by the trainers enhanced the successful implementation of projects and programmes at all levels of the organization. It incorporates participatory approaches in developing and implementing projects at community and national levels.

The organization's members are engaged in a process where they collect data on the impact of climate change and other development challenges affecting their communities and country. The members of CYEN worked with residents of the communities to analyse the data collected and then organize a meeting with the wider community to discuss the issues evolving from the data analysis. They also developed concept project proposals to resolve the environmental challenges confronting the community (Leach, 1996; Chambers, 2002; Simmons, 2018, 2021).

This stage of the project implementation process consists of a team from the organization working with the community to develop project proposals to submit to donor agencies to access funding for the project. The organization would then work with the community to implement the environmental project. During the project implementation process, the CYEN team would work with the local community to enhance their capacity, including (a) developing the planning processes at the community level; (b) building the monitoring, evaluation and reporting systems; (c) developing transparent budgetary systems; and (d) sharing of the lessons learned and best practices from the project with other organizations and communities. The bottom-up participatory approaches implemented by the organization are process-oriented and based on the notion of "learning together", indicating a female member of the organization.

The participatory development model developed by CYEN is based on the teaching of Paulo Freire, the Brazilian educator and philosopher. Conscientization is used by Paulo Freire to imply action and organization (Mics, 1973). The work of Paulo Freire is very relevant to the Caribbean as it empowers disadvantaged and marginalized people of northern Brazil to organize themselves and take action to resolve the development challenges affecting their communities. Freire was cognizant that one could not change the existing power structure by making people simply aware of the power structure. He states that a cultural revolution cannot substitute for a political revolution and further holds the view that a meagre modification of the power structure without a cultural revolution, i.e. without a heightened awareness by the people, would again repeat the old hierarchical system of the domination of man and women over the others. Freire, therefore, thinks that the cultural revolution, i.e. the education for the "new society", must begin in the old society. He contends that education could only achieve its true aim if it is part of a larger movement for liberation (Leach, 1996; Mics, 1973).

This conscientization model was developed and promoted by Paulo Freire, the CYEN bottom-up approach CYEN has utilized to work with vulnerable young people and their communities across the Caribbean. There is no short-term, quick, fixed or "instant-coffee" approach to tackling the devastating environmental and climate change challenges facing the people in small island states of the Caribbean. Due to the severity of the environmental challenges in the Caribbean, it would take a considerable amount of time to build the capacity of young people and their communities to respond adequately to such challenges. CYEN anticipates that there will be difficulties in mobilizing the masses to be part of the change movement. However, its decades of experience in the environmental field demonstrated that people would come on board in solidarity to adapt to climate change's impact and reduce the region's environmental degradation. Therefore, investing in young people is the best option for securing a climate-free future in the Caribbean.

Coaching and mentoring are effective methodologies employed by the organization to build the capacity of young people and their communities. Outside school members are encouraged to work with younger people still in schools, colleges and universities to enhance their skills and knowledge on the critical environmental issues threatening the Caribbean. This method also enables experienced members and experts to transfer knowledge, skills and strategies to the less experienced and knowledgeable members of the organization. Networking and partnership building are also important strategies used to build the capacity of young people on the environment and sustainable development issues. CYEN, a network of youth organizations operating at national and regional levels, encourages organizations to develop networks and partnerships with development organizations at national, regional and international levels. These networks facilitate sharing of skills, knowledge and technical and financial resources among partner agencies and affiliates to implement sustainable projects and programmes.

CYEN utilizes social media to reach numerous young people across the Caribbean with new knowledge and skills. It uses WhatsApp, Facebook, Zoom and other social media platforms to train young people. Social media platforms are much easier and cheaper to use than organizing physical workshops and conferences. CYEN collaborates with the Caribbean social media platform to implement its online and other capacity development activities. Although it has not been conducted any research to check the effectiveness of online training compared to holding physical workshops and conferences, it is of the view that workshops and conferences, even though they may be more expensive to organize, are more beneficial than online courses to members. Several young people who tuned into the workshops implemented by CYEN were not available to respond to the questions raised by the trainer when questions were asked to seek clarification on an issue. There is a belief that many people would link to these online training courses via their phone and laptop computers but would not participate in the actual training. They would use the opportunity to do other activities while the training is being held.

5.2.10 The Effectiveness of the Methods Used by the Organization in Its Capacity-Building Programme

CYEN utilizes several methods to deliver its capacity development programmes to young people and their communities.

> The methodologies used by CYEN were very effective, especially with the hands-on activities. The internet-based activities allowed persons to participate where/when they were unable to travel.
> Renee Boyce-Drakes, former chairman of CYEN.

A similar view was supported by Yoland London, National Youth Coordinator (St Vincent and the Grenadines) and CYEN executive member, who argues:

> The capacity-building methodologies implemented by CYP were very effective. The combination of participatory methodologies and online training provides opportunities for young people to engage in these training activities. Several young people when on to pursing studies in college and universities, leading to master's and PhD degrees because of the exposure to CYEN. These young people are now taking on leadership roles in governments, NGOs, academic institutions, and in the private sector.

Although the training methods utilized by CYEN were deemed to be very effective, there were concerns by the key informants regarding the effectiveness of the methodologies in enabling the organizations to achieve their outcomes. Reginald "Reggie" Burke, executive coordinator, CYEN, postulates:

> CYEN uses a combination of methodologies, including training workshops and conferences and online platforms, to deliver new knowledge and skills to build the capacity of young people in the Caribbean. These methodologies were very effective. However, there are concerns that the environmental situation in the regions is getting worst at an alarming rate.

All the informants who participated in the interview process attested to the effectiveness of the methodologies utilized by CYEN to deliver the capacity-building programme (Chanika Mataya et al., 2020). They argued that these capacity-building methodologies should be viewed as a package and should not be seen in isolation from each other. The informants alluded to the progress made by individual members and the national networks of CYEN over the years, e.g., former members are in critical positions in governments and decision-making work as senior civil servants and ministers of government.

The informants argued that the capacity-building methodologies utilized by CYEN motivated young people across the region to pursue studies in the environment and sustainable development fields and are now taking up influential roles in making a decision on the development and implementation of the national programme and programmes aimed at the achievements of the Sustainable Development Goals (SDGs). The informants also highlighted the peer-to-peer, networking and mentorship approaches as empowering and efficient capacity-building methodologies. These methodologies enable members to transfer knowledge and skills from one young person to the other and vice versa. The methodological approaches also build comradery and friendship among young people. However, one informant was sceptical about the effectiveness of some of the methods and programmes implemented by the organization. He expressed his frustration that despite all these dynamic methodologies, the time and resources invested in enhancing the capacity of young people, it seems like nothing has changed; things seem to remain the same, and, in some circumstances, the environmental situation in the Caribbean appears to be worsening. He argues that "we are still talking about all the environmental challenges we were talking about in the 1980s prior to the establishment of CYEN".

Three factors should be examined when responding to the above concerns of the key informants. These are (a) the persons who were youth in the 1990s are now adults. The state of being a youth is not constant; it is transitional. Therefore, youth should be viewed as a transition process, and no one stays as a youth for the entire period of their life. The young people would become adults and move out of the CYEN into the wider society, the same way that the young person will leave his/her parental home and establish a home for his/herself as indicated in the World Bank life cycle transitional approach (World Bank, 2005).

In the same way, there are new groups of children who are moving into youthhood and becoming members of CYEN, most of which do not have any knowledge and skills on the environmental challenges impacting the Caribbean (Simmons, 2006; Commonwealth Secretariat, 2007; World Bank, 2005), and (b) the environmental challenges affecting the Caribbean has worsened tremendously due to the severity of the impact of climate stressors on Small Island Developing States. The Caribbean region and other SIDS produced less than 1% of greenhouse gases which causes global warming, which is responsible for causing climate change but are suffering from the full impact of climate change (Nurse et al., 2014; IPCC 2014a, b, c, 2021; Simmons, 2021), and (c) the reckless and unsustainable environmental policies and programmes implemented by governments in Caribbean SIDS are creating severe degradation on the environment (Simmons, 2021; IPCC, 2022). Therefore, one can conclude that what has happened in the Caribbean has nothing to do with the effectiveness of the methods used by CYEN to deliver its capacity-building programmes.

5.2.11 Key Lessons

Overall, despite the scarcity of the limited financial resources and technical capacity available to the organization to implement its programmes, CYEN extracted several valuable lessons learned from its capacity-building programmes:

- CYEN was highly praised for its consistent messaging and engagement of young people in climate change adaptation programmes in the Caribbean. CYEN instilled in young people a sense of patience and hope to change the environmental situation in the Caribbean now and in the future. Apart from hurricanes and storm events, the impact of Climate stressors on the environment may not be immediate and visibly felt by everyone. It may take a long period for young people to experience the severity of the climate change impact on their islands. Through CYEN work on climate change, we were taught that some climate stressors would take a longer period than others to experience the devastating impact of climate change. CYEN's work demonstrated that one must be consistent, committed and dedicated to building people's resilience to adapt to the impact of climate change in the region. Therefore, to give up means death for many, while persistence gives hope and future life.
- Thc future impacts of climate change are challenging to predict; consequently, one must believe and convince that the work the organization is doing on climate change adaptation is for the greater good of humanity. The legacy of CYEN's work is that environmental activism is essential for defending the natural resource base of Caribbean SIDS, thus keeping it on track for generations to come.
- CYEN has built a monitoring, evaluation and reporting (MER) system to provide feedback to donors, beneficiaries, partner agencies and other stakeholders on the performance of the projects/programme implemented by the organization. Although it was late, the MER assisted CYEN in documenting its work to ensure a sustainable future.
- The work of CYEN demonstrates a sense of persistence, commitment and confidence toward the achievement of Sustainable Development Goals in the Caribbean. This level of persistence on the part of the organization convinces donor agencies to provide funding to support the organization's work. One female youth leader said that working on environmental issues is like "walking down a lonely road alone". CYEN is prepared to walk down that road alone, hoping partner agencies will receive the message and join the organization on that important journey.
- CYEN observed that increases in awareness and education do not lead to actions and positive change in the environment and climate change adaptation. Therefore, there is an urgent need for a re-examination of the strategies, approaches, the type of programmes implemented and outcomes anticipated from the project/programmes implemented by the organization. This leads to a complete rethink of how the organization approach the building of the resilience of young people to adapt to the impact of climate change and the achievement of the SDGs.

5.2.12 Best Practices

Over the past two decades, CYEN developed an excellent track record of working with young people to enhance their capacity to adapt to climate change and the sustainable development challenges affecting the Caribbean region.

Numerous best practices are evolving from CYEN's work regarding the strategies implemented to resolve the region's environmental issues.

CYEN strategic programmes are designed to meet the needs of young persons at national and regional levels. This allows the organization to capture the interest of young people and then engage them in capacity-building programmes to enhance their awareness of the magnitude of the impact and allows them to participate in activities which contribute to the development of their community and nation.

- The bottom-up approach utilized by CYEN is viable for empowering and mobilizing young people and their organizations/communities to participate in environmental and sustainable development initiatives. The approach values their contribution to achieving the organization's and their country's outcomes at community, national and regional levels.
- The participatory approach is used by CYEN to enhance the quality of the project management process. It empowers young people and other stakeholders to participate in the project development by building partnerships among all stakeholders.
- The sharing of success stories and best practices on youth climate change adaptation and environmental conservation practices with groups, communities, the Caribbean region and at the global level leads to the building of confidence in the organization's membership to deliver more successful projects at community and national levels.
- The organization's success largely depends on the quality of training and capacity-building strategies provided to the organization's membership. These strategies equip members with the appropriate skills and knowledge of the environment, which benefits the organization, communities and the personal development of the members. The organization adapts to the practice of a learning organization, i.e. it is open to new and innovative learning practices and is willing to adapt and adopt new knowledge, skills and techniques evolving from implementing new learning practices.
- It is committed to the concept of "each one teaches one". CYEN encourages sharing knowledge and skills through the peer-to-peer learning approach and is a firm practitioner of the holistic growth and development of its members and their communities. Representation of the organization is not confined to the persons in the organization's leadership. CYEN allows every member to represent the organization at national, regional and international levels. This is also extended to a member of the communities in which the organization implements its programme.
- CYEN takes a holistic, balanced, cooperative approach to tackling environmental and sustainable development issues across the Caribbean region. The people deemed the natural resource end users should be part of the emerging management model implemented to resolve the environmental challenges in the Caribbean region.
- The structure of the CYEN consists of a regional executive supported by a national focal point responsible for the management and coordination of the national country chapters. The organizational structure supports the smooth coordination of the organization's programmes within member countries and across the Caribbean. An organization's governance and decision-making processes are based on a bottom-up approach involving countries setting their own agenda on the environmental issues they wish to work on. These national programmes are then utilized to shape the regional programme priority areas of the organization.
- Having older professionals working with CYEN regional office in Barbados encourages donors, the private sector and governmental agencies to provide funding to support the implementation of projects and programmes. The professional experts provide coaching and mentorship support to young people in the organization. These activities facilitate the transfer of knowledge and skills to young people in the organization, and the strategies

enhance donor agencies' confidence in funding the organization's programmes.

5.2.13 Recommendations

CYEN is one of the most progressive environmental and development organizations in the Caribbean. Therefore, the organization needs to continuously put measures in place to continuously enhance the capacity of its members to sustain the status of the organization as the leading regional development organization working with young people on the environment and sustainable development issues. It is hoped that the recommendations below will be implemented to build the organization's capacity.

- Create an enabling environment for young people to maximize their talents so they feel more empowered and encouraged to participate in deciding the type of development they want to implement at the community and national levels.
- Mobilize adequate finances to support the implementation of its programmes. The organization faces a severe challenge to mobilize sufficient finances to fund environmental programmes. CYEN frequently receives requests from communities to implement projects but would lack the financial resources to implement and support such projects. It should consider lobbying regional governments to offer subventions and special assistance to fund the organization's administrative and operational costs.
- To enhance the capacity of the organization at the national level, governments should explore the possibility of (a) providing office space for the national network to function and (b) sending staff from the public service to provide technical assistance and support to the CYEN national network to develop strategic plans and prepare project proposal for mobilizing financial resources.
- On the issues of securing sustainable finance for CYEN, the government should provide letters of support to the organization to donors such as the Global Environment Facility (GEF), the Green Climate Fund (GCF) and other multilateral and bilateral agencies to fund the environment and climate change adaptation projects implemented by CYEN. CYEN should work with governments and the corporate community to establish a youth environment trust fund to finance national and regional environmental and sustainable development projects.
- The government in the Caribbean should support policy initiatives to include young people in delegations attending international meetings, including the Commonwealth Heads of Government Meeting, UNFCCC and other high-level meetings/conferences.
- The governments can write projects and mobilize funding to support environmental and climate change projects. The government should work with CYEN to implement joint capacity-building workshops on project writing and management, strategic planning and other critical areas of sustainable development. It should explore the possibilities of providing internship and apprenticeship opportunities to youth in ministries to facilitate the transfer of knowledge and skills to young people. CYEN should also explore capacity-building opportunities for young people in organizations within the private sector and bilateral and multilateral agencies.
- CYEN should focus its resources on developing programmes to support the engagement and participation of young people, especially young men, in sustainable development at national and regional levels.

5.3 Case Study on Capacity-Building for RAY, Northern Uganda (1997–2013)

The third section of the chapter explores the capacity-building work implemented by the NRS RAY project in Northern Uganda. The Norwegian Refugee Council (NRC) initiated the project, working in partnership with several developmental organizations to address the refugee crisis brought about by the civil war in Northern Uganda. The case study explores the strength, challenges, methodologies implemented and their effectiveness, the lesson learned and best practices and makes recommendations to enhance the implementation of similar capacity programmes in the future.

5.3.1 The Conflict and Refugee Crisis in Northern Uganda

The government of Uganda and the Lord's Resistance Army (LRA) were engaged in a bitter, prolonged armed conflict in Northern Uganda for over 24 years. The civil war resulted in a devastating humanitarian crisis. The conflict resulted in the displacement of approximately 1,800,000 people and the loss of tens of thousands of innocent lives. Internally displaced person (IDP) camps were established as homes for over one-eighth of the population of Uganda. The people lived in life-threatening conditions, with little access to water and sanitary facilities, sustainable livelihood opportunities, education and other basic social facilities and essential services (Barnes et al., 2013).

The effects of the conflict were extremely harsh on young people, children and women. Uganda has a young population, with approximately 70% of the population being between the ages of 16 and 24. Unemployed and underemployment are severe development challenges affecting over 50% of displaced young people in Northern Uganda. A similar number of young people are not involved in education, skills training and/or decent employment (ILO, 2010; Barnes et al., 2013; UNDP, 2015).

The conflict in Northern Uganda has made it extremely difficult for children and youth to participate in the formal education system. The educational opportunities beyond the fourth grade of primary school are not accessible to most of the young population. In the instance where it is accessible to education, there is evidence of (a) poor curriculum and instruction in educational content and (b) a mismatch between the skills provided by education and training institutions and those required by the market (Carter, 2008; ILO, 2010). This situation resulted in young people who do not have the knowledge and skills necessary to enable them to create their own employment and/or effectively work with others. The conflict contributed to increases in poverty, resulting in children and young people dropping out of school due to the displacement and poverty caused by the conflict.

There is also the absence of any active private sector and/or legal business activities in the area, resulting in a scarcity of jobs and a youth unemployment rate that exceeds 50% of the population. The civil war led to the destruction and/or closure of the business operated in Northern Uganda, which also contributed to the collapse of the agriculture and tourism sectors in the area (Barnes et al., 2013).

The incidence of illiteracy is very high among the population in Northern Uganda. Approximately 51% of women and 15% of men in Northern Uganda cannot read or write. Across Uganda, 23% of females aged 15–24 and 21% of males aged 15–24 are classified as functional illiterate, i.e. they lack the literacy and numeracy skills to enable them to fill out forms, write applications for a job and submit written complaints to their parliamentarian and/or local government officials on issues that affects their communities (Uganda Bureau of Statistics (UBOS) and ICF International Inc., 2012). These illiterate young people cannot sufficiently contribute to developing and reconstructing a post-conflict society.

The conflict brutally impacted the lives and well-being of children in Northern Uganda. It is estimated that over 25,000 children were kidnapped and fervently drafted as child soldiers to fight in the warring factions of the opposing armies. It is critical to note that children constituted 80% of the Lord's Resistance Army (LRA) rebel force. In cases where persons who were captured by the rebels tried to escape from the rebel confinements, the children were forced to kill their own parents, relatives and friends. The situation worsened when girl children were given away to commanders of the warring factions as wives and sex slaves and eventually sexually abused by these old men (Barnes et al., 2013).

The concerns relating to the poor education situation in Uganda are linked to the absence of functioning education providers and socio-economic cultural practices in Northern Uganda. Over a third of the poor household are headed by females and do not participate in the formal education system. The civil wars and conflicts attributed to the mass dislocation of people in the northern areas of Uganda. Such a situation resulted in the severe curtailment of the rights of children and youth to education.

5.3.2 The Norwegian Refugee Council (NRC) in Uganda

In 1997 the Norwegian Refugee Council (NRC) decided to participate in Uganda's humanitarian and development crisis. This decision was taken due to the rapid displacement of the population in Acholi land triggered by the civil war in Northern Uganda. Over the past decade, the institution provided protection and humanitarian assistance to refugees and internally displaced persons from its offices located in Gulu and Kitgum. After signing a peace agreement among warring parties, refugees left the camp to return to their original area of residence. However, the prolonged civil war destroyed most educational, health and housing facilities.

The NRC devised a strategic action plan with supporting programmes to contribute to Northern Uganda's recovery by implementing a robust educational project.

The overall aim of the project was to protect and promote the rights of the displaced people in humanitarian need by improving living conditions and seeking durable solutions. The education programme sought to contribute to the acceleration of educational attainment and sustainable livelihoods of the Acholi area by increasing access to and quality of education by:

(a) Improving the availability and quality of basic school infrastructure, including classrooms, classroom furniture and sanitation facilities in return areas.
(b) Improving the availability (attraction, retention and attendance) of teachers in hard-to-reach return areas through the construction of teachers' houses.
(c) Improving teachers' competency and commitment through teacher training and allowances.
(d) Increasing participation of refugees by:

- Subsidizing the household's financial burden of children's schooling.
- Implementing non-formal alternative basic schooling and skills training approaches to mitigate the effect of widespread demand for children and youth labour in household livelihood strategies.

5.3.3 The Recovery of Acholi Youth (RAY) Project

The situation in Acholi continued to be fragile and unpredictable when the RAY project began in 2011. Those displaced, especially young people, were suspicious about instituting livelihood projects in return areas after years of conflict and violence exploding in these communities.

NRC Uganda secured a 3-year funding grant from the Norwegian Ministry of Foreign Affairs through the Royal Embassy of Norway in Kampala to implement the Recovery for Acholi Youth project. The project's purpose was to preserve access to basic quality and appropriate education, i.e. enhanced knowledge and skills training facilities for out-of-school children and youth for future self-reliance, economic recovery, sustainable peace and stability in the Acholi sub-region of Northern Uganda. The project intended to enable out-of-school children and youth in return areas to access and complete the basic primary education and skills training cycle. The RAY project was scheduled to end in December 2013.

The RAY project had three components supported by three basic strategies and interventions:

1. The Accelerated Learning Programme (ALP) enables out-of-school children in return areas to access basic education.
2. Youth Education Pack (YEP) empowers out-of-school youth in return areas to access basic literacy and numeracy, life skills and occupational skills training.
3. School construction (SC) focuses on improving the availability of basic school infrastructure in return areas.

The RAY project was implemented in the seven districts of the Acholi land sub-region. The project impacted young people between 15 and 24 years for the Youth Education Pack and children between 9 and 14 years for ALP. As a core element of the RAY exit strategy, the project was designed to be incorporated with government educational priorities to ensure the project's sustainability in Northern Uganda. The project was implemented on a phase-in/phase-out basis, including (a) the school construction component phased out at the end of 2011, (b) the YEP component phased out at the end of 2012 and (c) the ALP programme component was closed with the rest of the country office at the end 2013.

5.3.4 Project Aim and Objectives of the Project

NCR has worked in Northern Uganda, providing protection and humanitarian assistance to refugees and internally displaced persons. Much of NRC's work in Northern Uganda was characterized by working within a volatile dysfunctional environment within an unpredictable, diverse movement of displaced people looking for security and enhanced well-being (Barnes et al., 2013).

The RAY education programme promoted and hastened the voluntary return of refugees and displaced people in Northern Uganda. The project's aim is to improve the knowledge, skills and overall competencies of displaced youth and children through their exposure and participation in quality education by:

- Improving the availability and quality of safe and secure learning environments, including classrooms, classroom furniture and sanitation facilities in return areas.
- Improving the availability (attraction, retention and attendance) of teachers in hard-to-reach return areas through the construction of teachers' houses.
- Improving teachers' competency and commitment through teacher training and allowances.
- Increasing participation by subsidizing the households' financial burden of children's schooling.
- Implementing non-formal alternative basic schooling and skills training approaches to mitigate the effect of widespread demand for children and youth labour in household livelihood strategies.

In its 3 years, the project was designed to address some challenges of disrupted basic education for children and young people in the Acholi sub-region. The project was intended to contribute to the wider recovery of the region and, overall, was a cornerstone of NRC's strategy for Northern Uganda.

5.3.5 Methodologies Implemented to Collect Data for the Project

This section of the book builds on the methodology section in Chap. 2. A range of qualitative and quantitative tools was utilized to collect data on the project. These tools include semi-structured interviews, questionnaires and focus group discussions (FGDs). Data were collected from reviewing project reports, journal articles, newspaper articles and other documentation on the project (Robson, 2002; Chamber, 2002; Yin, 2003; Angelsen et al., 2011; Barnes et al., 2013).

The semi-structured interviews were conducted with senior officials of the project, governmental officials and the heads of schools. These data were recorded using hand-written notes detailing all the points made in the interview. Notes for interviews and FGDs were transcribed and were qualitatively codified using NVIVO 11 to extract and compare themes. Questionnaires were conducted with students to draw a picture of their learning journey, including their future aspirations. Similarly, questionnaires were utilized to collect data from teachers and instructors in the skills programme.

A community contracting methodological approach to project management was adhered to by NRC as a strategy to enhance local livelihoods and community economic development by employing community members and former YEP graduates in constructing these facilities. The approach utilizes the local resources based on the community residents and youths by providing opportunities for (a) graduates to practise their newly acquired skills alongside seasoned carpenters and masons, thus fostering respect and comradery and bridging the generation gap; (b) contributing to the development of the economic and infrastructural base of the community; (c) enabling ex-combatants from both sides of the conflict to work together for the development of the community; this strengthened opportunities for fostering peace and sustainable development (Barnes et al., 2013); and (d) providing opportunities for using the local raw material to construct environmentally friendly houses for teachers, classrooms and sanitary facilities.

Triangulation and Challenges in Data Analysis

The study used quantitative and qualitative data sources and methodologies to ensure triangulation (Robson, 2002). Emerging findings were drawn from this process by the evaluation team. These findings were tested through three processes: (a) comparison of findings and conclusions from previous evaluations of ALP and YEP, (b) comparison to the international literature on non-formal education in post-conflict contexts and (c) presentation to the NRC country team, followed by discussion and further exploration of the issues.

ALP Sampling

From the sample frame, a random sample of ALP communities was made and classified by the district. Data collection was undertaken in all districts for ALP. The schools were selected based on the number in each district. Districts with more than six schools had four selected; those with less than six schools had three selected. The total number of schools included in the evaluation was 24. This is equivalent to 60% of the schools where the interventions occurred. The evaluators worked with the school authorities to select approximately 16 current ALP students from each school.

5.3.6 Components of the RAY Capacity-Building Programmes

Recovery of Acholi Youth *(RAY)* project was intended to ensure access to basic quality and relevant education and skills training for out-of-school children and youth. The project's purpose was to enable out-of-school children and youth in return areas to access and complete the basic primary education and skills training cycle (Barnes et al., 2013).

The project has the following components:

Accelerated Learning Programme (ALP) enables out-of-school children to access basic education in return areas. The ALP programme is viewed as a bridging initiative combining accelerated learning with "catch-up" objectives and is designed to support out-of-school children to complete their primary education and/or transition into the mainstream formal education system or YEP. The programme was mainly implemented at existing primary schools within the sub-districts. Four ALP centres (3 in Amuru and 1 in Nwoya) were standalone because these areas had no schools. One centre in the Gulu district was an annexe to the host school. The head teachers at the host primary schools were in charge of overseeing the ALP initiative. These head teachers were provided additional training and a monthly incentive for the additional work assigned (Barnes et al., 2013).

Youth Education Pack (YEP) was designed to enhance social inclusion by providing opportunities for young people in the Acholi subregions to acquire sustainable livelihoods through participation in a basic vocational skills training programme. The component enables out-of-school youth in return areas to access basic literacy and numeracy, life skills and occupational skills training. The YEP programme comprised four components: literacy/numeracy skills (basic reading and writing in the local language and English), life skills (including psychosocial support and entrepreneurship), basic vocational skills and income-generating activities in small support groups (once the learners have graduated and received a start-up kit). YEP targeted formerly abducted children, females, child mothers, single parents, orphans, child heads of households, youths without any previous formal primary schooling and ALP dropouts. The YEP attained a representation of 50% girls.

School construction (SC) improves the availability of basic school infrastructure in return areas. The SC programme consisted of building 70 houses for teachers in the 6 districts and constructing extra classrooms at schools implementing the ALP programme. Two five-stance toilet blocks (separate for males and females) were constructed per school. Teachers' houses did not include separate toilets. As noted above, YEP was based on existing buildings that were rehabilitated.

5.3.7 The Strengths of the RAY Project

From all accounts, the RAY project was designed and implemented to ensure access of out-of-school children and youth to basic quality education, and skills training in Northern Uganda was a tremendous success. The project's purpose was to enable out-of-school children and youth in return areas in Northern Uganda to access and complete their basic primary education and skills training. The success of the project was possible due to many strengths of the project which include the following.

ALP targeted qualified unemployed teachers, some of whom were not registered as teachers with the Ministry of Education. Although some local governments have committed to integrating the ALP teachers into the formal education system, the RAY project constantly competes for teachers with the formal primary system during implementation. As NRC and Echo Bravo recruited and trained new teachers, others moved to the formal school system because of the security of tenure.

A community contracting methodological approach to project management was adhered to by NRC as a strategy to contribute to enhancing local livelihoods and community economic development by employing members of the community and former YEP graduates to construct the classrooms, school toilets, housing for teachers and the renovation of buildings for the skills training facilities.

During the implementation of the RAY project, there were numerous instances where sub-counties, NGOs and the Church of Uganda have taken over the management and implementation of YEP centres since 2012. One of the sub-counties has remodelled the YEP centre into a secondary school. Therefore, on completion of the project in 2013, most of the centres/training facilities were either incorporated and managed by the government's Ministry of Education system or by the Church of Uganda and other NGOs.

The leadership of NRC provides opportunities for local authorities at the district level to participate in the management of the projects by providing regular information on the progress in implementing the programme. Districts were given opportunities to shape and influence the implementation process and the programme outcomes and engaged in the training of teachers/instructors for the YEP.

The NRC allowed the local authorities (districts and sub-districts) and the Ministry of Education to participate in the decision-making process. The location of the YEP centres at the community level was determined on the advice of the district education officers. Initially, the location of the centres was based mainly on the availability of permanent buildings and not on the studies of the distribution of young people. However, the RAY project targeted different locations from the previous YEP project. RAY thus benefited from the previous experience with YEP in Uganda.

The YEP instructors voiced their happiness with the "good management" and close guidance of the YEP centre by NRC. They viewed this as the main contribution to the high performance of young people after the training. This was accomplished since the NRC kept on its YEP programme staffs after the project's closure to provide technical advice and support to the graduates of the centres.

The NRC provided housing for teachers in schools to (a) ensure that they are in the community to provide direct support to learners when they need help and (b) facilitate the recruitment of teachers from communities outside of the district to work in remote rural and urban communities. The construction of houses for teachers positively impacted teacher attendance and retention, thereby enhancing the quality and efficiency in the delivery of education within the district. This endeavour also offered opportunities for parents to meet and work with teachers to enhance their children's education. It contributed to building cooperation and partnership between parents and teachers by helping to make them more aware of their roles and responsibilities in the education process (Barnes et al., 2013).

5.3.8 Weaknesses of the RAY Project

Like most humanitarian projects implemented in developing countries, the RAY project has many weaknesses. The NRC was in the process of implementing the coding of ALP community schools and teachers' houses for them to become government-aided/assisted teachers. However, after the withdrawal of direct budget support by donors, the district education officer (DEO) was instructed to halt this process. After building houses for teachers, NRC wanted to add solar panels to enable the teachers' houses to produce electricity; however, they were discouraged by the DEOs, who wanted to avoid setting high standards that the government could not maintain. Stakeholders raised some concerns about the skills aspects of the curriculum offered in the YEP package. Based on market assessments using the Women's Refugee Commission tool, some trades were dropped, and others were added due to the demand for skills required by the market.

The YEP package included literacy and numeracy, life skills, business skills, basic agricultural production, income generation and a group start-up kit for graduation. Stakeholders expressed their concerns about the programme not offering courses in vocations more related to the agriculture value chain since this occupation is the predominant economic activity in the area. Although agriculture is an important livelihood activity in Uganda, the young people rejected the option of agriculture with greater interest in urban-related trades. This trend is shared with other post-conflict countries such as Liberia and Sierra Leone. While Northern Uganda has its own distinctive farming calendar. This farming calendar conflicts with the national education programme's calendar. The farming calendar was found to interrupt the YEP calendar, resulting in a high level of absenteeism (45%) and eventual dropout of students who could not catch up with the education curriculum when they returned to school.

The teachers were paid the same salary as those working in the formal education system to enhance the project's long-term sustainability. Yet, the consistency of NRC teachers being paid on time was not enough to complete with remuneration package offered by the government. There was a mass resignation of teachers, resulting in a drop in salaries offered by the organizations that took over YEP centres. Therefore, it is critical to ensure that local organizations desirous of taking over such facilities from external donor agencies possess the financial and technical capacity to sustain such projects for an acceptable period.

Although teachers hired in the ALP were graduates of government-accredited institutions and were scrutinized by the district education officers, there were concerns from stakeholders that some ALP teachers were not utilizing modern teaching methods for training learners. Thus, some learners did not grasp the concepts. The education officials were also concerned with the poor quality of training provided to teachers working on the ALP component of the project. They concluded that professional teachers who can assess individual learners' needs and create individual work plans for the learners to accelerate through the ALP are needed in the project.

5.3.9 Challenges Identified in Enhancing the Capacity of the RAY

NRC faced challenges in implementing such humanitarian and educational projects; the basic infrastructure in the area, such as roads, water and electricity, was completely damaged during the civil war. There was the absence of classrooms, toilet facilities, houses for teachers and facilities to run the skills training centre. This was compounded by the lack of human capacities, such as teachers and instructors, to teach the children in schools and skills training centres due to the mass displacement of people in Northern Uganda.

Stakeholders raised concerns about the skills aspects of the curriculum offered in the YEP package. These market assessments using the Women's Refugee Commission tool show a mismatch pertaining to the skills offered by the skills training centre and those required by the market. The NCR and the Ministry of Education have decided to terminate some of the courses offered to students and add more relevant skills training courses to meet the market's needs.

Education officials also voiced concerns regarding the quality of training offered by teachers to students participating in the ALP component of the project. These education officials were not satisfied with the quality of the training of the teachers working on the ALP component of the project. They insisted that (a) the project needs to recruit professional teachers who can assess the learner's needs and create individual work plans for each learner to accelerate through the ALP and (b) the training provided to teachers should put more emphasis on training teachers' skills to evaluate learning needs and develop a work plan for individual students.

Although agriculture is an important livelihood activity in Northern Uganda, the young people rejected being trained in agriculture and instead showed a greater preference for training in the trades. This trend is shared with other post-conflict countries such as Liberia and Sierra Leone. Due to the rural base nature of Northern Uganda, many students had to leave schools and skills training centres for a substantial amount of time to work with their families to grow their crops on the farm. On returning to the classroom, these students would find it difficult to catch up on their classwork and drop out of classes. Students felt that future training in agriculture would further consign them to a miserable existence in the rural communities of Northern Uganda.

5.3.10 Capacity-Building Problems and Strategies Implemented to Address These Challenges

The civil war causes mass displacement of people in Northern Uganda. Due to the ferocity of the war, poor security of the area, the remoteness and inadequate infrastructure in these communities, professional workers such as teachers flee to more peaceful urban centres in search of peace, security and sustainable livelihoods. The project encountered difficulties recruiting trained teachers to teach students in the classroom and manage the skills training centres. NRC implemented the following strategies to resolve the capacity-building challenges: (a) it provides a stipend to head teachers working in schools in the area to manage the acceleration of learning and skills training centres, (b) the construction of teaching houses to encourage teachers to stay in the area and (c) the training of persons in accelerated teacher training programmes to prepare them to teach students in schools and skills training centre.

The absence and/or scarcity of skilled builders such as carpenters, masons and electricians to construct the class classroom, school toilets and teacher houses and renovate the buildings destroyed by the war to run the skills training centres were of severe concern for the implementors of the project. The NRC hired seasoned skilled trade persons from the area to work with graduates from the skills training centres and former combatants from both sides of the conflict to work on the project. The graduates from the skilled training centres were exposed to new sustainable construction techniques and skills in building constructions which adhere to more vigilant environmental standards and climate-friendly practices.

The strategy contributed to achieving these outcomes by:

(a) Building peace, friendship and comradery among former combatants.
(b) Enhancing the economy of the area and that of individual households to lead to stability and peace.

The staffs of skills training centres were concerned with the narrowed core skills offered to students, considering the skills required by the local market. This resulted in a mismatch in the courses offered by the skills training centre and those required by the market (Carter, 2008; ILO, 2010). This prompted the NCR to offer more skills courses at the skills training centres which enhanced the participation of women in traditionally male-dominated skills such as brick-concrete masonry and carpentry-joinery. The centres also ran a cooling-off and transfer period which provided opportunities for students to sample other trades and choose to swap if they wanted to acquire another trade. This strategy introduces learners to the potential income benefits from market-orientated agriculture, sustainable tourism and entrepreneurship. While the curriculum was important to learners, they also highlighted the social aspects of YEP centres and relationships with the instructors as important variables in their educational process. Another capacity-building problem is where the farm calendar clashes with the education calendars resulting in students dropping out of classes due to the considerable amount of time they spent farming on their family plots.

5.3.11 Methodologies Utilized and Their Effectiveness in Enhancing Capacity-Building

To confront the shortfall in recruiting teachers to teach in the classroom and manage skills training centres, ALP targeted qualified unemployed teachers, some of whom were not registered as teachers with the Ministry of Education. Although some local government authorities have committed to integrating the ALP teachers into the formal education system, the RAY project constantly competes for teachers with the formal primary system. To resolve these challenges, the NRC and Echo Bravo recruited and trained new teachers through accelerated accredit teachers training programmes. There were constant vacancies in the Ministry of Education for trained teachers, so teachers working on the project would move to the formal education system for employment because of better security of tenure.

The teacher training courses provided by NRC utilized a variety of methodologies to deliver teaching content, knowledge and skills to trainees. These methodologies include lectures, workshops, participatory group exercises, peer-to-peer training activities, mentorship, coaching and internship. The trainee teachers were sent on teaching practice exercises to understudy the work of experienced professional teachers. These professional teachers would mentor and coach the student teacher to facilitate the transfer of new knowledge and skill in education. Occasionally, a team of external educators from NRC, the local government authorities and the Ministry of Education would observe the student teacher's work. The team of visiting educators would provide feedback to the trainee teacher to enhance the performance. These methodologies were deemed highly effective by the teachers and NRC officials.

The instructors of the skills training centres indicated that they would start their course by introducing the background and career opportunities to students before providing instruction. The survey data reveals that this approach was followed-up by providing opportunities for students to practise their respective skills in the classroom. The learners were instructed to protectively use the tools and equipment to ensure adherence to health and safety rules and standards. They were then given specific projects/assignments to work on individually or in a group setting. Working in groups is encouraged by instructors and teachers to transform these groups into small businesses and enterprises (Simmons, 2022).

The learners' view on the quality of the teaching methods administered by teachers and instructors in the RAY project is excellent. Approximately 95% of the learners said they were happy with the quality of the teaching methods used by their teachers and instructors. Of that amount, 74% said they were very happy with the teaching methods that other teachers and instructors implemented. The level of classroom teaching and interactions was also quite high. Over three-quarters (77%) of the learners said they received 4–6 periods or lessons of actual teaching per day. They accorded high ratings as a result of the following methods used in teaching reading: (a) listening to someone reading aloud, (b) silent reading, (c) learning new vocabulary from a text, (d) pronouncing or sounding words, (e) reading comprehension, (f) taking home books to read and (g) reading other materials at home (Barnes et al., 2013; Chanika Mataya et al., 2020).

A community contracting methodological approach to project management was adhered to by NCR. This bottom-up methodological approach was designed by NRC as a strategy to enhance local livelihoods and community economic development in Northern Uganda (Matland, 1995; OECD, 2013: Barnes et al., 2013; Butler et al., 2015). From all accounts, this approach was highly influential in building peace, comradery and friendship among ex-combatants and the rest of the community (Barnes et al., 2013).

5.3.12 Lesson Learned

The lesson learned from the RAY capacity-building project in Northern Uganda (Barnes et al., 2013) are listed as follows:

- Offering psychosocial support training for teachers to enable them to address issues of brutalization among learners and tackle violent, disruptive behaviour without resorting to corporal punishment.
- Employ qualified teachers to ensure that the right people are used to deliver the right content and skills to learners. ALP and YEP are concepts that formerly trained teachers can comprehend easily with additional training. Such actions would save the project valuable time and resources during the early stages.
- Implementing refresher and orientation training courses and workshops for teachers and instructors to enhance the delivery of quality education provisions, including feedback.
- When schools and skills training facilities do not have their own water source, getting clean and safe water for the pupils/learners while in school becomes difficult, and it takes a lot of time for learners to fetch their own water to bring to the schools. Such valuable time should have been better spent on furthering their education. Schools and skills training facilities should have their own water supplies on the school compound to ensure that children/learners dedicate the time to improve their education.
- Some communities in Northern Uganda are so disadvantaged that clean and safe water becomes part and parcel of any intervention. Therefore, ensuring that clean water is supplied to these communities is the first step toward providing quality, sustainable education.
- Early recruitment, training and deployment of teachers in the communities where the educa-

tional facilities are located must be considered before starting the schools. The teacher should complete their teacher training before posting them to work in the schools and skills training facilities.

- Constructing teacher housing would normally enhance the willingness of teachers to work in remote rural communities. It is important to note that qualified teachers would easily be willing to teach even in locations far away from their homes once house accommodations are provided as part of the package.
- Joint monitoring by key stakeholders and feedback are crucial as it minimizes teacher absenteeism and ensures prompt actions. Organizing pupil-teacher associations (PTAs) and other support community education groups to work with teachers to support the delivery of sustained quality education in remote and marginalized rural and urban communities is essential.
- The high levels of poverty in Northern Uganda were caused by the massive displacement of people resulting from the civil war. The ALP was successful because the full costs of implementing the education provision were met.
- The parents do not have the financial resources to afford the cost of educating their children nor is the government in any position to cover such cost. Therefore, the government must mobilize donor support to fund such projects in the future until the government is in a better financial position to support such educational initiatives.
- Participation in competitive co-curricular activities such as sports and physical education would only be possible if efforts were made to support the host schools with game/sports equipment and teacher training.
- Establishing ALP centres in host schools and collaborating with the host school administration enable the ALP learners to benefit from a range of sports and physical education activities within the host school.
- The provision of sports equipment should be part of the support the organization provides, and there should also be monitoring of usage and participation of learners in sports. The provision of school uniforms for sports/games should be part of the support package offered to schools participating in the project.
- Unemployment and poverty among young people in Northern Uganda are very high, and the RAY project must focus on the creation of sustainable livelihoods and employment opportunities among the youth sectors as a strategy to reduce the migration of young people, women and men from the area to more lucrative urban centres and other countries.
- Start-up toolkits are necessary when providing vocational skills training to poor youth because they cannot take off in their trades studied with the knowledge acquired only. They need a basic tool to find work and/or create their own businesses.
- Deliberate campaigns to other stakeholders to support YEP graduates by offering them contracts for certain works, tax waivers and continuous follow-ups are of utmost necessity. Constant monitoring of groups after they have established their businesses helps them improve their business's quality and overall well-being.
- Market assessments must be thoroughly undertaken to indicate whether the trades established by RAY graduates can offer income-generating opportunities.

5.3.13 Best Practices

The best practice evolving from the NRC RAY capacity-building project:

- The NRC has developed a viable model of partnership and cooperation among the technical skilled educators and officials from the local government authorities and the Ministry of Education in Northern Uganda to offer support to teachers. This strategy was essential to raise the credibility of non-formal education and provide a long-term support base for the education sub-sector.
- The coordinating and management of exchange visits, mentorship and coaching of teachers by senior professional teachers at

schools that are role models contributed to the facilitation and exchange of ideas, the provision of an enhanced sense of community and the sharing of problem-solving theories and practice. The close observation and supervision of teachers during their training contributed to the quality of knowledge imparted during the teacher training, providing feedback and enhancing the learning experience.

- The RAY demonstrated the necessity of engaging all stakeholders, especially the Ministry of Education, from the beginning of the project development process to ensure the acceptance and integration of the project outcome in Uganda's national development planning processes. It is critical to identify the specific roles each stakeholder will place in the partnership, e.g. the government's education officials possess the knowledge, skills and competencies in terms of the recruitment, training, monitoring and supervision of the teachers. Therefore, NRC must collaborate with local government officials at the district level to recruit teachers to work on educational projects.
- Providing safe, clean water supply, sanitary facilities and other environmental considerations are critical to the RAY education project. Although these provisions were not part of the initial project, NRC established a framework of collaboration and partnership with the Ministry of Education and local government authorities to ensure that such provisions are made to enhance water, public health and sanitary conditions in communities where schools and technical skills training centre are located.
- The NCR instituted a mechanism to involve groups of learners in decision-making on issues that encourage their learning and well-being outside the school environment. This practice makes it much easier to empower learners to participate independently in activities which enhance their self-confidence, self-worth and well-being. The learners involved in the skills training programmes can work on projects such as building furniture, houses, fixing motor vehicles, etc., to generate income through practising their craft/art/skills.

5.3.14 Recommendations

NRC has successfully implemented the RAY project in Northern Uganda through the implementation of a series of capacity-building programmes, including (a) the training and recruitment of teachers to work with displaced children and youth, (b) the setting up of skills training centres to train youth in sustainable livelihoods skills and (c) the recruitment of graduates of skills training programmes to work alongside skilled builders to construct teachers' house, toilets and classrooms/training facilities. Despite these achievements, there were uncertainties around the future of the ALP centres and the funding of the education of children and youth from marginalized communities in Northern Uganda. The morale of learners and parents was affected because they did not see any likelihood of them staying in education if the material provided by NRC was terminated (Barnes et al., 2013).

- NRC field staff must work with the local education authority (LEA) to encourage learners and parents to complete the entire year of their studies by championing ALP to the communities. It is hoped that government and stakeholders will inject funding to support the continuation of the project.
- NRC must renegotiate the content and process of ALP with communities before the termination of the project. It is required for NRC to use the remaining period to revisit the ALP design with communities, school management committees and parent-teacher associations to assess the gaps in the design and project implementation process and make recommendations for integrating the project into the formal education system. It is essential to secure the legacy of ALP by developing a trust fund to support poor and marginalized children and youth to continue their education to the sec-

ondary level and their participation in advanced skills training programmes which leads to the creation of their own businesses and sustainable livelihoods.

- High demand still exists for the implementation of the ALP in rural areas. This project approach is still viable as (a) many areas have not yet benefitted from ALP and (b) work is to be done on enhancing the quality of education offered in these communities. Additionally, the government has demonstrated a willingness to employ some of the NRC teachers in the formal classroom. To protect the legacy of ALP, the government should mobilize resources for upscaling the RAY into a national programme to enhance deficiencies in the quality of education provided to citizens.
- Develop a special strategy for child mothers in education programming. The evaluation found that the vulnerable group most likely to drop out of RAY or to struggle with attendance was child mothers – particularly heads of households. While providing baby care in education facilities helped, this did not remove the burdens of domestic responsibilities or unsupportive families. A special strategy is needed to consider child mothers at the design stage, outside of the context of the school or skills training centre settings.
- Introduce start-up kits and groups into YEP earlier in the project development cycle. Start-up kits are an important incentive for learners and are critical to motivating them to establish their own businesses. The graduates have no relationship with these toolkits by giving YEP start-up kits to learners at the end of their training. They do not feel the tools are theirs and often discard them as soon as they complete their studies.
- Developing employment and enterprise groups is also important in any YEP programme. These groups need to be formed earlier in training to develop the group dynamics and build the necessary partnership and cooperation between prospective business partners.
- Pursue innovative approaches to reducing the costs associated with providing educational buildings. Building construction is a major cost driver in both ALP and YEP. There was an attempt to mitigate this by using existing structures and host schools. But, even so, a large part of the RAY budget was invested in structures that cannot be moved if needs change. This could explore options for flexible education buildings that could be disassembled and moved to other communities and districts if necessary.
- Extend the impact of education programmes through local-level partnerships with businesses and other NGOs. One YEP centre in RAY was set up and run with a partner, Pajule, from the beginning of the project. This was a highly positive experience and makes a case for more local partnerships to be established at the design and inception stages.

5.3.15 Summary of Conclusion

The chapter covers case studies on (a) the Commonwealth Youth Programme (CYP), (b) Caribbean Youth Environment Network (CYEN) and (c) the Recovery of Acholi Youth (RAY) project in Northern Uganda in efforts to build the capacity of young people to tackle the education, environment and sustainable development challenges confronting these institutions, communities, countries and regions. The case studies explore the strength, weaknesses/challenges, methodologies implemented and their effectiveness, the lesson learned, the best practices and present recommendations to enhance the capacity-building initiatives of the institutions in the future.

The first case study analyses the capacity-building initiatives implemented by the Commonwealth Youth Programme (CYP), the Youth Affairs Division of the Commonwealth Secretariat based in London, UK. The organization works internationally with governments and youth NGOs in 54 countries in Europe, Africa, Asia, the Caribbean and the Pacific regions to build the capacity of governmental institutions and youth NGOs utilizing a hybrid (top-down and bottom-up) approach to enhance good governance, peace, the rule of law and SDGs in member states of the Commonwealth (Matland, 1995;

OECD, 2013). Although the organization has its own weaknesses, such as the lack of a sustainable funding mechanism to implement its programmes in the field and the absence of dedicated staffs to focus on its capacity-building and research programmes, generally, the capacity-building methodologies implemented by the organization during the period under review were very effective in building the capacity of young people and their organization across the Commonwealth. Due to its regional presence through the existence of the regional centre in the four designated regions of the Commonwealth, it was able to implement tailor-made capacity development programmes which fit the development challenges affecting the young people and the governments in the region.

The second case study examines the effectiveness of the CYEN's capacity-building initiatives implemented in 15 countries in the Caribbean region. CYEN is a regional youth environment and development network established by young people 30 years ago to enhance the knowledge, skills and actions of young people and their organizations to protect the environment and achieve Sustainable Development Goals. CYEN encounters challenges such as the absence of a sustainable funding source to support the implementation of its projects. It has effectively implemented a bottom-up approach which empowers young people to voice their concerns on the development issues affecting their community, country and region and provides opportunities for young people to participate in national and regional decision-making processes on the impact of climate change and other development calamities (OECD, 2013; Matland, 1995). CYEN has also developed a track record of working with young people to enhance their capacity to adapt to climate change and sustainable development challenges in the Caribbean region.

The third case study explores the capacity-building strategies implemented during the Recovery of Acholi Youth (RAY) project. The initial project to rescue the displaced people and refugees in Northern Uganda was designed as a top-down project implemented by the government of Uganda; however, as NRC and other partners came on board, the approach shifted to utilizing a hybrid approach (Matland, 1995; OECD, 2013). RAY was implemented by the Norwegian Refugee Council (NRC) in partnership with the government of Uganda in Northern Uganda. The initial project was to rescue the refugees and displaced people fleeing the devastation of the civil war in Northern Uganda. The project brought about several critical partners working in collaboration to resolve the refugee crisis in Northern Uganda. The component of the project that is being assessed was funded by the Norwegian Refugee Council (NRC), working in collaboration with civil society organizations and the local educational authorities and the Ministry of Education. From all accounts, the RAY project, designed and implemented to ensure access of out-of-school children and youth to basic quality education and skills training in Northern Uganda, was a tremendous success.

The RAY project was designed to enable out-of-school children and youth in return areas in Northern Uganda to access and complete their basic primary education and skills training. The teacher training courses provided by NRC utilized participatory bottom-up methodologies to deliver teaching content, knowledge and skills to trainees. These methodologies include lectures, workshops, participatory group exercises, peer-to-peer training activities, mentorship, coaching and internship. The trainee teachers were sent on teaching practice exercises to understudy the work of experienced professional teachers. Professional teachers work as mentors and coach the trainee teacher to facilitate the transfer of new knowledge and skill effectively and efficiently in the education system.

Although two of the case studies presented, the exclusion of CYEN, were not directly environment and climate change adaption projects, however, all the cases presented lessons learned and best practices that could be used by academics, policymakers and development practitioners to guide the implementation of a capacity-building project in climate change adaption and sustainable development. The cases conclude with a recommendation for enhancing similar projects in the future.

References

Angelsen, A., Larsen, H. O., Lund, J. F., Smith-Hall, C., & Wunder, S. (2011). *Measuring livelihoods and environmental dependence; methods for research and fieldwork*. Earthscan.

Barnes, J., Simmons, A., Abola, C., Nkawanzi, V., Semay, S., & Greeley, S. (2013, November 25). *Norwegian refugee council recovery of acholi youth independent evaluation report*. Impact ready; Version 2.4.

Bloomfield, G., Bucht, K., Martínez-Hernández, J. C., Ramírez-Soto, A. F., Cheseña-Hernández, I., Lucio-Palacio, C. R., & Inzunza, E. R. (2017). Capacity-building to advance the United Nations sustainable development goals: An overview of tools and approaches related to sustainable land management. *Journal of Sustainable Forestry*. https://doi.org/10.1080/10549811.2017.1359097

Butler, J. R. A., Wise, R. M., Skewes, T. D., Bohensky, E. L., Peterson, N., Suadnya, W., Yanuartati, Y., Handayani, T., Habibi, P., Puspadi, K., Bou, N., Vaghelo, D., & Rochester, W. (2015). Integrating top-down and bottom-up adaptation planning to build adaptive capacity: A structured learning approach. *Coastal Management, 43*, 346–364. https://doi.org/10.1080/08920753.2015.1046802; Taylor & Francis Group, LLC; ISSN: 0892-0753 print/1521-0421 online.

CARIBSAVE. (2010). *Climate change risk profile for St. Vincent and the Grenadines*. CARIBSAVE.

CARIBSAVE. (2012). The CARIBSAVE: Climate change risk atlas (CCCRA): *Climate change risk profile for Saint Vincent and the Grenadines* (p. viii). Available at: http://www.caribbeanclimate.bz/closed-projects/2009-2011-the-caribsave-climate

Carter, R. (2008). *Caribbean youth, an integrated literature review*.

Chamber, R. (2002). *Participatory workshops: A sourcebook of 21 sets of ideas and activities* (Vol. 794, pp. 421–434). Earthscan.

Chanika Mataya, D., Vincent, K., & Dougill, A. J. (2020). How can we effectively build capacity to adapt to climate change? Insights from Malawi. *Climate and Development, 12*(9), 781–790. https://doi.org/10.1080/17565529.2019.1694480. Francis and Taylor Group.

Charles, H. (2006). *The Commonwealth response to global challenges facing the 21st century* (Vol. 4, No. 2, p. 49). Commonwealth Youth and Development, Department of Commonwealth Studies, UNISA (University of South Africa Press).

Commonwealth Secretariat. (2007). *The commonwealth plan of action on youth empowerment 2007–2015*. Commonwealth Youth Programme, Commonwealth Secretariat.

Commonwealth Secretariat. (2021, August 9). *New Global Youth Development Index shows improvement in the state of young people: Global Youth Development Index Global Scores*. Commonwealth Secretariat London. https://thecommonwealth.org/news/new-global-youth-development-index-shows-improvement-state-young-people

Gomez, O. (2013). *Climate change and migration: A review of the literature*. International Institute of the Social Studies/Erasmus University Rotterdam.

https://www.un.org/ohrlls/content/about-small-island--developing-states#:~:text=Small%20Island%20Developing%20States%20(SIDS,social%2C%20economic%20and%20environmental%20vulnerabilities

https://www.unep.org/

ILO. (2010). *A skilled workforce for sustainable growth: A G20 training strategy*. International Labour Office.

ILO. (2011). *Greening the global economy: Skills challenge*. Skills for employment policy brief.

IPCC. (2014a). Climate change (2014): Synthesis report. In *Contribution of Working Groups I, II and III to the Fifth Assessment Report of the Intergovernmental Panel on Climate change* (p. 151). IPCC.

IPCC. (2014b). Summary for policymakers. Climate change impacts, adaptation and vulnerability. In C. B. Barros, V. R. Dokken, D. Mach, K. J. Mastrandrea, M. D. Bilir, T. E. Chatterjee, M. Ebi, & K. L. Estrada (Eds.), *Contributions of the Working Group II to the Fifth Assessment Report, Field*.

IPCC. (2014c). Climate change: Synthesis report. In *Contribution of Working Group 1, 11, 111 to the Fifth Assessment Report of the Intergovernmental Panel on Climate change* (p. 151). IPCC.

IPCC (2022) Climate change 2022 Mitigation of Climate change, *Summary for Policy Makers: Working Group 111 Contribution to the Sixth Assessment Report of the International Panel on Climate change* (AR6 WG 111).

Joseph, T. (2006). *National Youth Policy: Case studies from St Lucia and Trinidad and Tobago* (Vol. 4, No. 2). Department of Commonwealth Studies, UNISA (University of South Africa Press).[1]

Leach, M. (1996). *A revolution of ideas and action: Interview with Andrew Simmons*. Share International. July/August 1998 Issues of Share International. https://

- 1 *Key Informants for CYP*Dwynette Eversley, Programme Officer of CYP Regional Centre, Caribbean.
- Elizabeth Namwala, Chipata community. Zambia.
- Florida Harrison of Chitekwere Group in Malawi.
- Judith Webb, Associate Professor, UTT and CYP Development Consultant.
- Dr. Henry Charles, Acting Director, Commonwealth Youth Affairs Division, Commonwealth Secretariat.

- *Key Informants for CYEN*Renee Boyce-Drakes, former Chairman of CYEN.
- Reginald "Reggie" Burke, Executive Coordinator, CYEN.
- Yoland London, National Youth Coordinator (St Vincent and the Grenadines) and CYEN Executive Member.

www.share-international.org/archieves/environmental/en_mlrevolution.htm. Download 5 Sept 2022.

Matland, R. (1995). Synthesising the implementation literature: The ambiguity-conflict model of policy implementation. *Journal of Public Administration Research and Theory, 5*(2), 145–174.

Maunders, D. (2006). *Youth work train and methodologies* (Vol. 4, No. 2). Department of Commonwealth Studies, UNISA (University of South Africa Press).

Mics, M. (1973, September 29). Paulo Freire's Method of Education: Conscientisation in Latin America. *Economic and Political Weekly, 8*(39), 1764–1767. (4 pages) Published By: Economic and Political Weekly; https://www.jstor.org/stable/4363063

Mishra, P. (2007). Emigration and brain drain: Evidence from the Caribbean. *The B.E. Journal of Economic Analysis & Policy, 7*(1), 1547–1547. https://doi.org/10.2202/1935-1682.1547

Nurse, L. A., McLean, R. F., Agard, J., Briguglio, L. P., Duvat-Magnan, V., Pelesikoti, N., Tompkins, E., & Webb, A. (2014). Small islands. In V. R. Barros, C. B. Field, D. J. Dokken, M. D. Mastrandrea, K. J. Mach, T. E. Bilir, M. Chatterjee, K. L. Ebi, Y. O. Estrada, R. C. Genova, B. Girma, E. S. Kissel, A. N. Levy, S. MacCracken, P. R. Mastrandrea, & L. L. White (Eds.), *Climate change 2014: Impacts, adaptation, and vulnerability. Part B: Regional aspects. Contribution of Working Group II to the Fifth Assessment Report of the Intergovernmental Panel on Climate change* (pp. 1613–1654). Cambridge University Press.

OECD. (2013). *The nature of policy change and implementation: A review of different theoretical approaches*. Organisation for Economic Cooperation and Development.

Robson, C. (2002). *Real world research, a resource for social scientists and practitioner researchers*. Blackwell Publishers.

Ryan, C. (2006). *Youth enterprises and sustainable enterprises* (Vol. 4, No. 2). Department of Commonwealth Studies, UNISA (University of South Africa Press).

Simmons, A. (2006). *The commonwealth youth development index: Challenges and prospects for youth* (Vol. 4, No. 2). Commonwealth Youth and Development, Department of Commonwealth Studies, UNISA (University of South Africa Press).

Simmons, A. (2018). *Impact of climate change on youth in small island communities: The case of St. Vincent and the Grenadines*. This thesis is submitted to De Montfort University in partial fulfilment of the requirements for the award of Doctor of Philosophy, School of Engineering and Sustainable Development De Montfort University.

Simmons, A. (2021). *Impact of climate change on youth in small Island developing communities* (Sustainable development series). Springer Nature.

Simmons, A. (2022). *Climate change adaptation framework and youth entrepreneurship in West Africa* (Sustainable development goals series). Springer Nature.

Stachowiak, S., Robles, L., Habtermanion, E., & Maltry, M. (2016). *Beyond win: Pathway for policy implementation*. ORS Impact/The Atlas Learning Project.

Takawira, I., & Simmons, A. (2004). Youth development at the Centre of the Commonwealth Agenda: Commonwealth Youth Programme (CYP): Forum 21. *European Journal on Youth Policy No. 3 12/2004*.

The World Bank annual report 2007 (English). Washington, D.C.: World Bank Group. http://documents.worldbank.org/curated/en/732761468779449524/The-World-Bank-annual-report-2007

Uganda Bureau of Statistics (UBOS) and ICF International Inc. (2012). *Uganda demographic and health survey 2011*. UBOS/ICF International Inc.

UN (1981) *Secretary-General's report to the general assembly*. A/36/215. https://documents-dds-ny.un.org/doc/UNDOC/GEN/N81/165/83/PDF/N8116583.pdf?OpenElement

UNDP. (2015). *Human development report*. Work for Human Development, United National Development Programme (UNDP).

UNEP (2013) *Identifying emerging issues from the perspective of Small Island Developing States (SIDS)*. Expert Group Meeting Co-hosted by United Nations Environment Program (UNEP) and the United Nations Department of Economic and Social Affairs (DESA), 14-16 May 2013, Cambridge, United Kingdom.

UNESCO. (2014). *Shaping the future we want: UN decade of education for sustainable development (2004–2014)*. DESD Monitoring and Evaluation.

White, B. (2012). Agriculture and the generation problem: Rural youth, employment, and future farming. *Young People and Agriculture in Africa; Institute for Development Studies; IDS Bulletin, 42*(6).

World Bank. (2005). *Children and youth: A framework for action* (pp. 7–21). World Bank.

Yin, R. K. (2003). *Case study research: Design and methods* (3rd ed.). Sage.

6 Reflections and Guidance for Implementing Capacity-Building Climate Resilience Programmes

Abstract

Chapter 6 presents a summary of the conclusion, outcomes and recommendations, including the implementation process, the deliverables and proposed actions for enhancing the institutional capacity of organizations to deliver climate change and sustainable development programmes at national, regional and international levels. The chapter is divided into seven sections. The sections included (a) a summary of the climate change stressors and drivers impacting the people and their communities; (b) a Caribbean perspective, an analysis of bottom-up and top-down approaches utilized to implement capacity-building programmes in St Vincent and the Grenadines; (c) a West African perspective, an analysis of bottom-up and top-down approaches utilized to implement capacity-building programmes in Sierra Leone; (d) an international perspective, an analysis of bottom-up and top-down approaches utilized to implement capacity-building programmes including the Commonwealth Youth Programme (CYP), the Caribbean Youth Environment Network (CYEN) and the Recovery of Acholia Youth (RAY) project in Northern Uganda; (e) relevance of the conceptual framework for capacity-building in developing countries (LDC and SIDS); (f) the limitation and pertinence of the book; and (g) a summary of lesson Learned, best practice and recommendations evolving from the case studies.

Keywords

West Africa · Caribbean · Northern Uganda · Caribbean Youth Environmental Network · Least developing countries · Best practice · Climate change adaptation · Sierra Leone · Bottom-up approach · Top-down approach · Capacity-building · Drivers · Stressor

This chapter presents a summary of the conclusion, outcomes and recommendations, including the implementation process, the deliverables and proposed actions for enhancing the institutional capacity of the organizations to deliver climate change adaptation (CCA) and sustainable development programmes at national, regional and international levels. The chapter is divided into seven sections. The sections included (a) a summary of the climate change stressors and drivers impacting the people and their communities; (b) a Caribbean perspective, an analysis of bottom-up and top-down approaches utilized to implement capacity-building programmes in St Vincent and the Grenadines; (c) a West African perspective, an analysis of bottom-up and top-down approaches utilized to implement capacity-building programmes in Sierra Leone; (d) an

A. Simmons, A.-A. Simmons, *Building Institutions for Resilience*, Sustainable Development Goals Series, https://doi.org/10.1007/978-3-031-28881-4_6

international perspective, an analysis of bottom-up and top-down approaches utilized to implement capacity-building programmes including the Commonwealth Youth Programme (CYP), the Caribbean Youth Environment Network (CYEN) and the Recovery of Acholia Youth (RAY) project in Northern Uganda; (e) relevance of the conceptual framework for capacity-building in developing countries (LDC and SIDS); (f) the limitation and pertinence of the book; and (g) a summary of lesson learned, best practice and recommendations evolving from the case studies.

6.1 Summary of the Climate Change Stressors and Drivers Impacting the People and Their Communities

Climate change severely impacts developing countries, specifically SIDS and LDCs (IPCC, 2014a, b, c, 2022, Nurse, 2014; Chatenoux & Wolf, 2013). It has adversely affected global natural ecosystems and human well-being (IPCC, 2014a, b, c, 2022). In the future, climate change can create considerable adverse effects throughout all geographical locations (IPCC 2014a, b, c, 2022). It has adversely affected local economies, livelihoods, health and well-being, specifically in developing countries (Hernandez-Delgado, 2015; Simmons, 2021, 2022). The climate change stressors impacting developing countries include storms, sea-level rise (SLR), temperature rise, drought, invasive species and human interference. These climatic pressures were verified through the interrogation of national reports, academic journals and other international reports on climate change. Field visits were conducted in Asia (south; 2015–2017), Africa (east and west; 2019–2022) and the Caribbean (2018–2021) to meet with key informers as well as to observe the severity of the impacts of climate change on the socio-economic and ecological systems. We also participated in national and regional conferences/meetings, including the UNFCCC COP 21 in Paris in 2015, to observe how policymakers, practitioners, youth, women, civil society, technicians and other key stakeholders are negotiating with their counterparts from developing countries to ensure a just and equitable society.

The appropriate institutional arrangements in LDC and SIDS are weak. Therefore, such institutional arrangements must be vigorously instituted to achieve the desirable outcomes (Adger et al., 2003; Ojwang et al., 2017). Such arrangements possess the substantial potential to decrease the impacts of climate change at all levels (Pahl-Wostl et al., 2013). The policymakers and officials from developing countries appeared to be less prepared than their counterparts at the climate change negotiations. Many delegations of policymakers and officials from developed countries were intimidating. The delegation from the entire Caribbean region was much smaller than those from developed countries, yet they must cover all the simultaneous parallel meetings and discussions. Although the limited member of delegates from the Caribbean organizes through Caribbean Community Centre for Climate Change (CCCCC) and the Organisation for the Eastern Caribbean States (OECS) Secretariat, I was generally satisfied with the outcomes of the Caribbean delegation. It was hard work keeping pace with the main events and ensuring that there were delegates to cover the hundreds of side events parallel to the main event. The Caribbean delegation would organize early morning briefing sessions when necessary to ensure that information on the outcomes of critical issues being negotiated is shared with delegates; delegates were allocated to cover the important proceedings schedule of the day. Similar strategic block and regional briefing meetings are arranged by delegations from developing African and Pacific countries.

Despite the level of strategic engagement on the part of delegates, one can conclude that some delegations lack the critical competencies and aptitude to negotiate at such international levels. This situation is compounded by a large proportion of policymakers from developing countries who cannot effectively negotiate on behalf of the people on climate change at the international level. Key informants believe that the poor representation at these international meetings is because of the ad hoc selection procedures of delegates in developing countries. In some developing countries, persons are selected to participate in high-level meetings due to (a) the seniority of the post of the officer rather than the competencies and capacity of the officials to add value and influence the outcomes of the meetings and (b) "repayment" and/or compensation to the official for completing a "job well done" for the political party in government and (c) if the officials are related to the minister and/or a high-level official of the government, and (d) some persons are appointed as diplomats if the individual has the financial resources to pay for the diplomatic seat, i.e. such position would give the individual access to such meetings.

A key informant argued that since the international body hosting the meetings would usually cover the cost of a minimum of two officials to attend these high-level meetings, the government officials who would be selected to attend such meetings would use the opportunities to participate in other political and economic meetings while they are in the city where the meetings are taking place. Due to the limited financial resources available in LDCs and SIDS, the governments would try to make full use of these opportunities by "piggybacking" on these meetings to achieve their political and economic outcomes. This area needs further exploration to assess the opportunity cost of such actions.

The government needs to ensure that the officials who are selected to participate in these high-level meetings (a) possess the competencies, skills and knowledge on the issues being discussed and (b) are mature enough to participate in the high-level discussion and empowered by the government/country to make the decision on the country's behalf. There are instances where government officials attend international meetings without the mandate from their capital to make decisions on behalf of the government. Substantial time is spent in meetings with senior officials from their capital to decide which position they should vote for. The meeting would move on to other agenda items before a decision is reached by their country's officials, thus losing out on an important opportunity.

Developing countries must invest time and resources in capacity-building initiatives to ensure a cadre of skilled policymakers and senior officials representing the country at international meetings and conferences (Bloomfield et al., 2017). Capacity-building and training are not only an important element for preparing senior officials and policymakers to attend international meetings and conferences but should be viewed as an essential feature of the preparation of the country to build the resilience of its people to adapt to the impact of climate change and the achievement of the Sustainable Development Goals (Bloomfield et al., 2017; https://www.uncdf.org/article/7637/building-capacity-an-essential-aspect-of-adaptation-action).

6.2 A Caribbean Perspective: An Analysis of Bottom-Up and Top-Down Approaches Utilized to Implement Capacity-Building Programmes in St Vincent and the Grenadines (Chap. 3)

The section of the book presents an analysis of the case studies on capacity-building development methodologies and initiatives utilized by the JEMS Progressive Community Organization and the Government Community Development Division (CDD), Ministry of Social Development, in St Vincent and the Grenadines (Chap. 3). One of the essential lessons from the case studies on JEMS and the Community Development Division is that community participation in consultative participatory development processes is essential to successfully implementing CCA and sustainable development programmes (Hernandez-Delgado, 2015; Jaja et al., 2016). The government of St Vincent and the Grenadines, Community Development Division's top-down approach to the implementation of community development programming failed miserably in St Vincent and the Grenadines due to (a) lack of participation of the beneficiaries in the design and development of the projects and (b) deficiency in skills and knowledge of CDD officials on strategies for engaging people at the community level in projects and programmes. The programmes of the CDD became successful when on recognition of the failure of the approach, a decision was made to shift to (a) the implementation of a hybrid approach and (b) engage in capacity-building to enhance the skills and knowledge of CDD officials on strategies for mobilizing the communities in the development of its programmes (Matland, 1995; OECD, 2006).

6.2.1 Strengths of the Capacity-Building Programmes

The CDD programme was funded through the British Overseas Division (DFID). Initially, there were few consultations with the people and their communities to assess the community's development challenges and develop viable programmes and projects to resolve the felt needs of the communities. These programmes were designed in the CDD headquarters in Kingstown, the capital of St Vincent and the Grenadines, by officials without consultation with the community. The projects were mainly implemented by the government's CD officers in the communities with limited participation of the people in implementing the projects. While comparatively JEMS, despite its limited funding, was implementing successful capacity-building projects utilizing bottom-up participatory development approaches to manage and implement community projects and programmes, at the same time, the government's community development programmes were failing. In many instances, the JEMS projects were funded by the government's CDD organization. During the period under review, JEMS successfully implemented pipe-borne water, climate disaster risk management, health and sanitation facilities, building sports and recreational facilities, literacy and continuing education, environment and conservation, agriculture, craft and entrepreneurship employment skills programmes in villages of Stubbs, King's Hill Forest Reserve, Enhams, Choppins, Carapan, Brighton, McCarthy, Argyle and Peruvian Vale on the southeast of St Vincent (Leach, 1996; Simmons, 2021).

The success of the JEMS community and sustainable development programmes was attributed to its bottom-up approach and capacity development programmes which train and empower residents to develop their leadership and participate in the decision-making process and resolve the development needs in their community. A considerable amount of JEMS resources were utilized to build residents' capacity using cultural wakes, popular theatre, drumming and calypso music to mobilize many people to participate in its community environment and sustainable development programmes (Leach, 1996; Simmons, 2021). Both the CDD and JEMS utilized capacity-building methodologies such as workshops, conferences, town hall meetings, internships, popular theatre and networks to build the capacity of the institutions, the people and the community to adapt to climate change and resolve the development challenges facing the communities (Bloomfield et al., 2017; Leach, 1996; Simmons, 2018). However, what was more important to note here is in recognition of the failure of the CDD programme, the government shifted to the implementation of a hybrid approach, which was accompanied by a) the recruitment of professional leadership for the ministry, d) the development and implementation of new strategic programmes and c) approval and implementation of new project appraisal and management facility that engage people and their communities (Matland, 1995; OECD, 2013; Stachowiak et al., 2016).

6.2.2 Weaknesses of the Capacity Development Programmes

Due to the government's top-down CDD programmes' failure, the donors raised concerns that the CDD programme was not reaching the marginalized people and their communities due to the direction the programme was moving. The government employed new leadership in the Ministry of Social Development and shifted its approach to a hybrid approach. The hybrid approach combines the strengths of top-down and bottom-up approaches. It engages people in the participatory process geared toward empowering them to identify the issues that affect their communities and develop strategies to resolve the felt needs in the community. The new approach provides the people with renewed momentum, confidence and a glimmer of hope resulting from the success of the programmes (Matland, 1995; OECD, 2013; Stachowiak et al., 2016). To build the adaptive capacity of communities, the new leadership of the CDD mainstreamed the participatory community development process across all government agencies and programmes as well as implemented a process of engaging the people by networking with civil society agencies (Armitage & Plummer, 2010; Smit & Wandel, 2006). Senior officials and policymakers in the government were trained in critical areas of development, including community engagement and sensitization, conservation and sustainable development. Officials of CDD were engaged in mentorship, evaluation, coaching, networking and other strategies to enhance their individual capacity and those of the community.

JEMS, as a community-based organization, has weaknesses which affect the capacity of the institutions, including (a) the lack of funding to support the implementation of its sustainable development programmes, (b) the constant challenge of migration of its members to the USA, Canada and other countries in the Caribbean in search of employment and enhanced well-being and c) increasing demands on the organization by the communities to respond to the needs of communities due to success of the organization's community sustainable development programme.

6.2.3 Capacity-Building Methodologies

The CDD implemented a programme of intense training and awareness building utilizing methodologies such as workshops, internships, apprenticeships and networking to build the leadership of the CDD and the communities by enhancing officers' knowledge and skills in innovative community sustainable development practices (Bloomfield et al., 2017). Other strategies employed by the division to enhance capacity include (a) mobilization of scholarships from the DFID for staff to pursue studies overseas; (b) implementation of quarterly training workshops for staff on various aspects of development; (c) organizing internships, exchanges and study tours both nationally and regionally for staff and community leaders to provide them with first-hand experiences on the implementation of sustainable development practices and programmes; and (d) provision of tuition fees and other educational provisions to support staff who lack the minimal requirement for employment in the civil service. These capacity-building initiatives were fully supported by the Ministry of Social Development and the head of the government's Public Service Commission (the training division).

Before the appointment of new leadership in the ministry in 1990, there was a breakdown in trust between communities and the officials at the government's CDD due to the unjustified utilization of financial and technical resources of the division. The community leaders accuse the government agency of lacking transparency and accountability in the project approval process. The new direction implemented by the government leads to the development of a new strategic participatory development approach to sustainable community development, which adheres to the gender mainstreaming approach to development. The new development approach to community development in St Vincent and the Grenadines was approved by the government's Cabinet, which approved a new project approval committee consisting of representatives from key governmental agencies, non-governmental organizations (NGOs) and other civil society institutions as well as from the private sector. These strategic actions contributed to improving the trust and transparency of the people in the CDD process and led to the creation of a dynamic and coherent sustainable development model.

The CDD built a partnership with multilateral and bilateral agencies such as DFID, OAS,

USAID and others to mobilize financial and technical resources to support the building of feeder roads, adult education and literacy programmes, day care, farmers' education, environment and sanitation facilities, agriculture and land management programmes and capacity-building initiatives in communities (Bloomfield et al., 2017; UN, 2015). CDD would endorse project proposals submitted by NGOs to donors to fund projects/programmes in communities. JEMS developed partnerships with bilateral and multilateral agencies, which led to the organization receiving funding from the UNDP GEF/SGP, CIDA, CCC and USAID to implement environmental and sustainable development programmes at community and national levels.

6.2.4 Challenges to Capacity-Building

The change in the government of St Vincent and the Grenadines in 2001 resulted in a shift in the direction of the CDD to a top-down development approach. The CDD no longer has the funding and technical experts to drive the development process in communities. The changes in governmental policies and the loss of aid funding from DFID mean that there was an absence of funding and technical resources to support community sustainable development actions within communities. The lack of funds in the CDD also affected the capacity-building work of civil society organizations such as JEMS to implement development projects to address the felt needs of the communities. Informants felt that, to some extent, the government is responsible for strangling the dynamic community spirit and volunteerism in St Vincent and the Grenadines. People would volunteer their time to work in their communities to build agricultural feeder roads, volunteer at children and youth summer camps, teach at literacy programmes, repair the school or clinic in their community and manage disaster emergency centres; however, whenever there is money available, the government will give contracts to their party supporters from outside of the community to do the work.

St Vincent and the Grenadines was endowed with some of the most innovative NGOs; however, most of the active NGOs, like JEMS, are either dormant or no longer in existence due to a) the leadership of the organizations migrated overseas to seek employment and/or b) the inability of the organization to mobilize new membership and financial resources to sustain the programmes of the organization. They felt that most countries in the Caribbean are experiencing similar challenges in mobilizing young people, specifically young men, to participate in youth and community programmes. This is an area where further research is needed to investigate the evidence to support such claims and assess the impact on achieving Sustainable Development Goals.

6.2.5 Lesson Learned and Good Practices

The key outcome from the case study of JEMS indicates that women and youths were deemed the most vulnerable to climate change and other development calamities impacting the communities. However, the most disadvantaged sectors of the population are readily mobilized to contribute to the development of the community. The magnitude of the involvement of women in community projects was identified as the main factor responsible for the successes of the community development and environmental projects implemented by CDD and JEMS. Men were attracted to participate in these projects mainly due to the large percentage of women working tirelessly as volunteers. The CDD and JEMS work emphasized the implementation of participatory development approaches such as cultural wake, capacity-building training and community mapping exercises, which are the main factors contributing to the increased participation of the people in the development of community activities. The participatory strategies give the people a sense of ownership, commitment and stake in the development of their community (Leach, 1996; Jaja et al., 2016).

Due to the innovative capacity-building initiatives implemented by both institutions, similar best practices were identified, including developing and implementing innovative community-integrated participatory approaches that contributed to building the community's capacity to enhance project management, monitoring, evaluation and reporting within participating communities and at national levels. Both JEMS and the CDD utilized the "cultural wakes" and aided "self-help" models to raise the awareness of the residents of the communities on the severity of the development challenges facing these communities and mobilized the people to participate in actions to enhance the infrastructure, skills and knowledge of residents.

Regarding the CDD programme, the shift to the hybrid approach provides the basis for the communities to work in partnership with the government and other external partners to mobilize funds and technical resources to implement basic needs projects in communities. The CDD facilitates the implementation of community development of programmes in communities by persuading donor agencies to provide financial and technical resources to resolve community-felt needs. It also advocates on behalf of communities for the central government to equitably distribute resources to develop communities (Matland, 1995; OECD, 2013; Stachowiak et al., 2016; Jaja et al., 2016).

6.3 A West African Perspective: An Analysis of Bottom-Up and Top-Down Approaches Utilized to Implement Capacity-Building Programmes in Sierra Leone (Chap. 4)

Section C of the chapter presents an analysis of the top-down approach implemented by the UNDP/government of Sierra Leone Youth Employment and Empowerment Project (YEEP) in 2012–2018 and the bottom-up approaches utilized by the Global Youth Network for Empowerment and Development Sierra Leone (GYNED-SL) to build the capacity of the institutions and key partner agencies to drive the youth development initiative in a post-conflict LDC. The YEEP centres on strengthening the capacity of the National Youth Commission (NAYCOM) and the Ministry of Youth Affairs (MoYA) by building the capacity of both institutions to deliver youth development programmes in Sierra Leone.

6.3.1 Strengths of the Institutional Capacity-Building Programmes

The YEEP consisted of the implementation of top-down development initiatives to build the capacity of the National Youth Commission (NAYCOM) and the Ministry of Youth Affairs to deliver innovative youth development programmes across the country. The outcomes of the YEEP achieved include (a) the development of Youth Development Partner Coordination Framework; (b) the preparation and implementation of the blueprint of the youth development programmes for Sierra Leone, 2014–2018; (c) the mapping of donor funding to support youth development programmes in Sierra Leone; (d) the provision of TVET training to enhance the livelihoods and the employability of youth in Sierra Leone; and (e) development of a mechanism for mainstreaming of youth development in government ministries and division programmes (ILO, 2010; Simmons, 2021, 2022; UNDP, 2015; University of the West Indies, 2011). The YEEP project worked with universities and colleges to create the career guidance centres, which trained counsellors to work with students to deliver career guidance training at the tertiary level. The project provided computer soft- and hardwares to tertiary institutions to expose students to modern technologies and content. During the period under review, thousands of students successfully participated in the training offered by these institutions.

The GYNED-SL is a civil society organization that focuses on implementing bottom-up approaches to build the capacity of young people and their communities to initiate sustainable development actions in Sierra Leone. The organization utilizes participatory approaches and practices to enhance entrepreneurship and the sustainable livelihoods of youth (Chamber, 2002; ILO, 2010, 2013; Poschen, 2015;Simmons, 2021, 2022 ; UNDP, 2015).

The outcomes of the UNDP/NAYCOM top-down approaches are summarized as follows: (a) the development of the Youth Coordination Development Framework, which provides a system for coordinating and mainstreaming all youth programmes/projects across governmental institutions and beyond. The framework was supported by the implementation of the M&E system, which provides feedback to the policymakers on the performance of the programme; (b) provides strategic directions to the government of Sierra Leone on the specific capacity-building methodologies to be implemented to enhance the capacity of the institutions including training of officials, teachers and instructors of colleges, universities and TVET centres, internship and apprenticeship opportunities for young people to enhance their employability and development of innovative strategies, policies and programmes; and (c) advices government on the specific policies and roadmaps to achieve different development outcomes relating to building the capacity of institutions and youth people to build the resilience of communities to resolve poverty and reduce the impact of climate change and other development calamities (Simmons, 2021, 2022; UNDP, 2015).

The strengths of the bottom-up approach implemented by GYNED-SL includes the following: (a) visible cutting-edge innovative youth development programmes; (b) recognition by government of Sierra Leone, development partners and the local community as the leading youth development organization in Sierra Leone working on youth employment and entrepreneurial sustainable livelihoods issues; (c) being staffed by a team of young professionals who are passionate and dedicated and volunteers committed to utilize entrepreneurship as a vehicle for young people to achieve sustainable livelihoods; (d) possessing good working relationship with MoYA and NAYCOM and other governmental agencies working with youth, women and other marginalized sectors of the population; (d) scope of work of the organization is national and focused and driven by a cadre of volunteers working in all districts utilizing innovative participatory development approaches to implement successful youth development programmes/projects in collaboration with MoYA, NAYCOM, UNDP, UNFPA, ADB, GIZ and other partners; (e) beneficiaries participated directly in making decisions on the nature of the project they wish to implement in their community; and (f) GYNED-SL helps young people to become volunteers within their communities by building their capacity through training, internship, mentoring and coaching (Bloomfield et al., 2017; Simmons, 2022).

6.3.2 Weaknesses of the Institutions/Programmes

These two case studies indicate their own set of weaknesses and idiosyncrasies. The UNDP/government of Sierra Leone YEEP project was criticized for not having enough local people to participate in the project's design, implementation and management. A key informant states that the strategies implemented were mainly driven by external experts and consultants to the detriment of the development of the expertise of local persons to work as consultants and development experts. The YEEP did not give much attention to engaging and developing the local capacities so they would be equipped to implement similar development projects in Sierra Leone. These challenges were attributed to the limited staff composition and competencies in NAYCOM and MoYA in relation to the task to be performed. The lack of adequate staff competencies limits the ability of the institution to build the capacity of partner institutions and communities in a meaningful and strategic manner. These problems were compounded by the complexities of the development challenges facing the institutions and the youth sector in Sierra Leone in a post-conflict environment.

The major weakness of the GYNED-SL bottom-up approach was its inability to mobilize much-needed funds to support the implementation of the programmes. The organization currently works in all districts in Sierra Leone. Due to the limited financial and human resources available to implement its programmes, management should decide to reduce the number of programmes implemented in communities and districts. This decision may enhance the organization's efficiency to better allocate and utilize its limited resources to achieve its outcomes. The organization also lacks experienced staff to effectively deliver the youth sustainable development livelihood programmes. It is encountering difficulties retaining experienced, qualified staff to work with due to the financial constraint it is experiencing (Horstmann & Schulz-Heiss, 2014). The organization relies too heavily on university volunteers and interns to sustain its programmes' implementation. Therefore, it should explore more innovative strategies to mobilize financial resources from donor agencies. It should develop a mechanism to maintain the organization's institutional memory, which could be lost due to high staff turnover.

6.3.3 Capacity-Building Methodologies and Their Effectiveness

Both institutions utilized similar capacity-building methodologies to enhance staff's skills, knowledge and competencies to deliver sustainable youth development programmes. The YEEP utilized training workshops as its lead strategy to build the capacity of staff, partner institutions and beneficiaries. The participants who attended these training workshops were provided with reference materials on various development topics to boost their competencies. Senior staffs of NAYCOM and MoYA were sent on regional and international training courses and meetings to learn how to utilize entrepreneurship to develop sustainable livelihoods within the youth sector. These training courses utilized methodologies such as whole-class presentations, panel discussions and group work activities to stimulate and sensitize beneficiaries and facilitate the transfer of new skills and knowledge (Bloomfield et al., 2017). GYNED-SL utilized a range of participatory and online methodologies to enhance the capacity of its staff, volunteers, youth and other beneficiaries. The organization set aside Mondays to implement its weekly staff internal capacity-building training. It would invite experts in relevant development fields to lead the training activities. The responses from the GYNED-SL staffs, volunteers and beneficiaries show that the blend of bottom-up participatory training methods implemented by the organization was very effective and contributed to enhancing the capacity of all beneficiaries (Bloomfield et al., 2017; Butler et al., 2015). Regarding the effectiveness of the capacity-building methodologies, a very small amount of UNDP/government of Sierra Leone YEEP officials believed that some of the methodologies were more effective than others.

6.3.4 Lessons Learned and Good Practices

The capacity development programmes implemented through the UNDP/government of Sierra Leone top-down YEEP and the GYNED-SL bottom-up capacity-building programmes enhanced the skills and knowledge of youth and officials to boost sustainable livelihood opportunities for the development of the youth sector. One of the key lessons learned from the Sierra Leone experience is that capacity-building works must be formulated and packaged strategically for them to be implanted in development institutions. This approach helped the institution sustain its capacities and memories when key personnel left the organization. It also contributes to harnessing the leadership and management capacity of other governmental agencies and partner institutions to achieve sustainable goals at local and national levels (Charles, 2006; Jayatilaka, 2003; Simmons, 2021, 2022).

The best practices evolved from the capacity-building programme focus on its contribution to enhancing the capacity of the staffs and beneficiaries of the organizations, key stakeholders and partners. The staffs of these agencies were also involved in all levels of the training and the consultation process and have acquired new skills and competencies resulting from their participation in the activities. The staff engagement enhanced the institutions' overall management and operation. Organizations can now better monitor, report, train, track and manage their development programme because of the capacity-building programmes initiated. However, the effectiveness and efficiency of these institutions in sustaining the development and implementation of the programmes have been constantly tested by the incidence of the Ebola pandemic, the devastating impacts of climate change, poverty, change of political leadership and other developmental calamities which impeded national growth and the achievement of the Sustainable Development Goals. Despite these development challenges, beneficiaries utilized the skills and knowledge acquired from the YEEP to deliver programmes to reduce the impact of the deadly Ebola epidemic, including performing roles of contact tracers, coordination, communication, awareness building and data management operation. These skills and knowledge contributed to enhancing the capacities of participating institutions to establish defensive remedial measures, which reduced the impact of the COVID-19 pandemic on the people of Sierra Leone. The YEEP also provided opportunities for beneficiaries to participate in the governance and management processes of the organization by creating fora for them to put forward ideas on the outcomes they wish to experience from their participation in the projects. The organization instituted mechanisms to ensure that it communicates effectively and efficiently with all stakeholders involved in the project development process. Although it was impossible to verify through the data collection process, informants stated that the young people who evolved from the YEEP process are now successfully running their own enterprises and effectively participating in decision-making at national, local authority and chieftain levels (ILO, 2010; UNDP, 2015).

6.4 An International Perspective: An Analysis of Bottom-Up and Top-Down Approaches Utilized to Implement Capacity-Building Programmes (Commonwealth Youth Programme, Caribbean Youth Environment Network (CYEN) and Recovery of Acholi Youth (RAY) Project in Northern Uganda) (Chap. 5)

This portion of the chapter presents an international perspective on the application of capacity-building. It offers an assessment of (a) the CYP, (b) CYEN and c) the RAY project in Northern Uganda to build the capacity of young people and their institutions to confront the development challenges facing these young people, their communities, countries and regions.

6.4.1 Top-Down, Bottom-Up and Hybrid Approaches to Capacity-Building

The CYP works internationally with governments and youth NGOs in 54 countries in Europe, Africa, Asia, the Caribbean and the Pacific regions to build the capacity of these institutions utilizing a hybrid (top-down and bottom-up) approach to enhance good governance, peace, the rule of law and SDGs in member states of the Commonwealth. Initially, being an intergovernmental organization, its normal approach would be to implement a top-down approach to developing its programmes across the Commonwealth. Nevertheless, realizing that this approach was not working, in 1997, the organization employed new staff in its London, UK, office who is experienced and qualified to implement participatory bottom-up approaches to enhance the management and implementation of its programmes. Thus, the organization's direction changes to a hybrid approach to developing and implementing its programmes (Matland, 1995; OECD, 2013; Stachowiak et al., 2016).

The CYEN is a regional civil society organization based in Barbados utilizing bottom-up approaches to build the capacity of young people and their communities to implement environmental and sustainable development programmes in the Caribbean region (https://cyen.org/). The third case explores the hybrid approach to capacity-building implemented by the Norwegian Refugee Council (NRC) as part of the RAY project. The RAY was part of a larger project developed to provide a safe "home" for refugees and displaced people fleeing from the devastation of the civil war in Northern Uganda (Barnes et al., 2013).

6.4.2 Strengths of Capacity-Building Programmes

The case studies possess varying and unique strengths. The CYP has a network of regional centres based in four regions of the Commonwealth, which developed and delivered "tailor-made" capacity-building programmes to governments and youth organizations to meet the specific need of the young people in member countries (Takawira & Simmons, 2004; Charles, 2006; Simmons, 2006). The staff of the CYP were selected from the regions of the Commonwealth where the CYP regional centre is based and are knowledgeable on the specific challenges facing the respective regions and the strategies to resolve the development challenges threatening the region. Through the Commonwealth Secretariat, CYP worked in partnership with international partner agencies, including the UN, World Bank and UNICEF, to mobilize resources to implement programmes to build the capacity of young people, member governments and stakeholder agencies to resolve their capacity-building challenges. Due to its regional presence Commonwealth Secretariat is able to implement tailor-made capacity-building programmes that fit and respond to the development challenges affecting the regional governments. These programmes utilized a range of methodologies, including training workshops, mentoring, coaching and networking (Chanika Mataya et al., 2020).

The strength of the RAY project was based on the capacity of the NRC to bring development partners on board to work in partnership with the government of Uganda to harness their technical and financial resources to resolve the refugee crisis brought about by the civil war in Northern Uganda. The project creates opportunities through (a) hiring and training of teachers and instructors to work in schools and skills training centres, (b) construction of classrooms for children and youth and construction of houses for teachers and (c) setting up of skills training centres for refugees and displaced youth and children to continue their education in a safe environment. The RAY project was designed and implemented to ensure access of out-of-school children and youth to basic quality education and skills training in Northern Uganda (Barnes et al., 2013).

CYEN is a regional youth environment and development network established by young people over 30 years ago to enhance the knowledge, skills and actions of young people and their organizations to protect the environment and achieve Sustainable Development Goals. The CYEN programmes empower young people to voice their concerns on development issues affecting their community and country and provide opportunities to participate in national, regional and international decision-making processes. Over the decades, CYEN successfully developed a track record for working with young people to enhance their capacity to adapt to climate change and resolve the sustainable development challenges in the Caribbean region (https://cyen.org/).

6.4.3 Weaknesses of Capacity-Building Programmes

Although the organizations implemented worthwhile capacity-building programmes to enhance the work of the groups and their beneficiaries, the funding to sustain their work in the field continues to be a significant challenge facing all the institutions/projects reviewed in the chapter (Dany et al., 2015). The CYP had dedicated funds committed by youth ministers on a tri-annual basis at the Commonwealth Youth Ministers Meeting (CYMM); however, due to the voluntary nature of the funding commitment, most governments would not pay to the youth fund, thus resulting in gaps in the funding of the programmes (https://thecommonwealth.org).

Funding has been a major challenge for CYEN since its establishment as an organization (Dany et al., 2015). The organization is constantly faced with the absence of a sustainable funding source to support the implementation of its projects. There is a tendency for donor agencies in the Caribbean to be cautious and sceptical regarding funding organizations managed by young people for fear of inappropriate use of project funds. The organization recruited older mature professionals to manage its office and to be the "face" of the organization. This has improved the confidence of donors to support the organization's regional and national programmes.

Although the RAY project in Northern Uganda has an approved funding source from the Norwegian government through the NRC, it has its own funding challenges. The RAY funding was committed to a specific project with a limited time frame and for a given purpose. After the RAY project was completed, the NRC, the Ministry of Education and the local authority encountered difficulties finding credible organizations with sustainable funding streams to manage the schools and skills training centres (Barnes et al., 2013). It was also realized during the implementation of the project that the water and sanitation facilities in the areas where the project was implemented were destroyed during the civil war, and there was no money allocated to repair these water supplies (Barnes et al., 2013).

6.4.4 Capacity-Building Methodologies and Their Effectiveness

The cases reviewed utilize a variety of methodologies to build the capacity of the organizations, partner agencies and the beneficiaries of the projects (Bloomfield et al., 2017; Chanika Mataya et al., 2020), namely, a) the purpose of the RAY was to enable out-of-school children and youth in return areas in Northern Uganda to access and complete their basic primary education and skills training cycle. Therefore, through training workshops, lectures, internships, participatory group exercises, peer-to-peer training activities, mentorship and coaching, the teacher and instructors' training courses enable the NRC to utilize different methods to deliver teaching content, knowledge and skills to trainees. The trainee teachers were sent on teaching practice exercises to understudy the work of experienced professional teachers. Professional teachers work as mentors and coaches to effectively transfer new education knowledge and skill to trainee teachers.

With the addition of online courses and networking, similar methodologies were used by CYEN and CYP to enhance the capacity of the organization and beneficiaries. CYP implemented formal and informal training programmes through the Diploma in Youth Work studies implemented by over 30 tertiary institutions across the Commonwealth. The capacity-building methodologies implemented by the organizations during the review period were very effective in building the capacity of governmental institutions, tertiary institutions, young people and their organizations. However, some informants expressed their concerns regarding the effectiveness of short-term training programmes over longer-term extended capacity-building programmes utilizing mixed methodologies (Chania Mataya et al., 2020).

6.4.5 Lessons Learned and Best Practices

Although two of the case studies presented (except for CYEN) were not directly related to the environment and climate change adaptation, all the cases present viable lessons learned and best practices that are relevant to be transferred to building the capacity of the organization working on climate change adaptation and sustainable development projects in developing countries. The lesson learned and best practices could be used by academics, policymakers and development practitioners to guide the implementation of capacity development projects in the future (check Chap. 5).

It is necessary to note that due to the uncertainty of funding of the CYP and the constant industrial actions of regional staff regarding pay disputes, a decision was taken a decade later by the Committee of Management of (COM) of CYP and the Commonwealth Secretariat to close the CYP regional centres in Africa (Zambia), Asia (India), Caribbean (Guyana) and the Pacific (Solomon Islands). Informants believe that the decision was premature, and more consideration should have been given to the issues by the Commonwealth Secretariat, especially in assessing the benefits the governments and young people are gaining from the regional presence of the CYP. The removal of the regional centre reduces the capacity of the CYP to consult with regional governments and key stakeholders across the Commonwealth on the critical issues affecting young people and to deliver "tailor-made" solutions to meet the specific needs of young people.

6.5 Review of the Application of the New Conceptual Framework for Building Capacity for Climate Change Adaptation in Developing Countries, the Limitation and Pertinence of the Booking, Including Best Practices and Lessons Learned

After closely examining the conceptual frameworks on building capacity for climate change adaptation presented in Chap. 2, we realized that the initial framework needed to be further reviewed to make it more appropriate for developing countries, specifically LDCs and SIDS.

6.5.1 Climate Change Adaptation Framework for Capacity-Building in Developing Countries

The updated framework consists of four steps; each is supported by a set of strategic tools to operationalize the framework. The steps of the *climate change adaptation framework for capacity-building in developing countries* are as follows:

Step 1: Climate risk vulnerability context

Step 1 is identical to step 1 in Chap. 2, as adapted from the DFID (1999 in Angelson, 2011), Sustainable Livelihood Framework (SLF) and the Youth Climate change Adaptation Framework (Simmons, 2018, 2021). The step frames vulnerability due to the impact of climate change stressors on people and their communities in developing countries. The step acts as an independent framework for how to apply theory into a practical perspective (DFID 1999 in Angelson, 2011; Simmons, 2021).

Step 2: The development of capacity-building strategies

Step 2 is also comparable to Chap. 2. The conceptual framework concentrates on the interpretation and analysis of the data collected on the nature and extent of the vulnerabilities affecting the community/country. It identifies and agrees on the appropriate capacity-building approach to responding to the vulnerabilities encountered by developing countries. A human capital approach was adapted to the conceptual framework as a strategy to build the adaptive capacity and resilience of communities and their institutions to adapt to the impact of climate change and other development calamities. This strategic approach ensures that human, financial and material resources are available to implement the community and national climate change adaptation plans and programmes successfully.

Step 3: Building capacity through linking strategic and operational plans with internal and external partners

Step 3 of the conceptual framework concentrates on implementing the adaptive capacity-building process and programmes at community and national levels utilizing top-down, bottom-up and/or hybrid approaches to capacity-building. This step is centred on implementing the community and national climate change adaptation plans and programmes and developing training and capacity-building curricula, manuals and methodologies. After carefully assessing the three stages, a decision was made to amend the stage to include monitoring, evaluation, accountability and inclusion related to the accountability perception from climate change-related treaties and agreements. Inclusion of learning is related to a) indigenous people and their technical knowledge, which are being practised worldwide, as well as b) the creation of learning organizations to implement capacity-building programmes and partner institutions as a strategy to capture best practices and lessons learned, particularly regarding climate change adaptation and mitigation. The approach also includes implementing participatory models for engaging key stakeholders (people, communities and institutions) and mobilizing human and financial resources (Horstmann and Schulz-Heiss, 2014; Callaghan, 2015). This strategy is important to ensure that adequate resources such as experts, materials and finances are mobilized to implement the adaptive climate change processes efficiently. This stage would provide guidance for the preparation of the Intended Nationally Determined Contribution (INDC) report for submission to UNFCCC COPs (Callaghan, 2015).

Step 4: Capacity development initiatives instituted

Step 4 is the same as in Chap. 2. The step centres on training and other capacity-building methodologies to raise awareness and engage the population in actions to build people's resilience to adapt to the impact and risk associated with climate change. Continuous education and training initiatives should be implemented to achieve the essential levels of capacity-building at the community and national levels (United Nations, 2015).

6.5.2 The Contribution of the Research to New Knowledge

The contribution of the research to new knowledge is assessed by concentrating on a) its relevance to building the capacity in developing countries, specifically LDCs and SIDS, b) institutional capacity-building, c) the theoretical contributions to new knowledge and d) the innovative methodologies and conceptual framework.

Relevance to Building the Capacity in Developing Countries, Namely, LDC and SIDS

Research shows that the rise in temperature, sea-level rise, drought and increase in storms can destroy people, countries and regions in developing countries (Chevallier, 2010; IPCC, 2014a, b, c, 2022). A report on climate change states that capacity-building is an essential element of the adaptation process; therefore, efforts to trace it separately are fundamentally problematic. The cross-cutting and integrated nature of capacity-building also hamper efforts to delineate the financial flows attached to the capacity-building component of projects. In this context, several reports underlined the need for an internationally agreed approach to quantitatively and/or qualitatively track capacity-building in official development assistance in general and in climate projects, especially when most developing countries indicated that capacity-building should be identified as the condition for implementing their Intended Nationally Determined Contributions (INDCs) under the Paris Agreement (Mizan Khan et al., 2019).

The developing country also noted that meeting/reporting responsibilities were difficult given their limited administrative and technical capacities (UNFCCC, 2021). Hence, the book submitted a conceptual framework for identifying the capacity-building needs of essential institutions and proposed a set of strategic approaches to redesign and better align national reporting guidelines to effectively respond to their needs to develop their capacity to enable them to build their resilience to adapt to the impact of climate change. The book benefits governments in developing countries by providing opportunities for policymakers and development practitioners to develop strategies and programmes aimed at building the capacity of communities and countries to enhance their resilience to adapt to climate change.

Governments can use the body of knowledge to implement CCA projects in local areas. This strategy would provide excellent opportunities for governments to be involved in more strategic activities such as evaluation, monitoring, accountability and learning by extracting lessons learned and best practices of the CCA project, mobilization of local and external funding and resources and the provision of technical assistance and support to local groups/organizations to implement CCA initiatives. By providing new and relevant information on capacity-building, the book would enable policymakers to create the necessary policies and enabling environment with the prerequisite skills and knowledge to build the resilience of the people to adapt to the impact of climate change.

Capacity-Building of the Institutions and Key Stakeholders

The devastating impacts of climate change severely affect the lives and well-being of people and their institutions, including their livelihoods and the natural ecosystems on which their livelihoods depend. Most developing countries in the south are faced with inadequate knowledge and skills to effectively build people's resilience to adapt to the impact of climate change. The book presents a new conceptual framework supported by innovative strategic approaches to build developing countries' capacity to adapt to climate change's impact.

Climate change adaptation is not scheduled high on the development agenda of most developing countries because it is not being viewed by politicians and the populace as an important issue. Most governments believe climate change is a long-term development issue whose impact will not be felt immediately until sometime in the distant future, a future they will not live to experience; thus, there is no need for them to spend the limited financial resources to address such issues when competing development issues warrant their full and immediate attention. There is empirical evidence concerning the lack of knowledge and skills by policymakers, practitioners and development planners and the wider population on the severity of the impact of climate change on the people and natural ecosystems in developing countries. This has resulted from a) the absence of a national environmental policy, b) limited allocation of technical and financial resources to build capacity and c) the lack of inclusion of climate change adaptation in the national educational curriculum of most developing countries.

Due to the severity of the impact of climate change in developing countries, policymakers, practitioners and other key stakeholders are working continuously to explore remedies to build the capacity of institutions to respond effectively to the impact of climate change. The book provides policymakers and practitioners with new skills and knowledge on mainstreaming capacity-building models in national policies and programmes. It will provide insights and learning on the unique contributions of institutions to building people's capacities to adapt to the impact of climate change and other development calamities in developing countries.

The Theoretical Contribution of the Research to Knowledge

The cases presented in this book underlie the need for adequate knowledge and skills to prepare policymakers and practitioners to address vulnerability, resilience and adaptation to climate change from a top-down and bottom-up perspective/approach. The research offers a set of strategies for enhancing the knowledge and skills of policymakers, students and practitioners on actions, approaches and practices to be implemented by organizations to enhance the capacity of the institutions involved in climate change adaptation and other development programmes, including top-down and bottom-up approaches in LDC and SIDS and other developing countries in Africa, Caribbean, Asia and the Pacific.

The book presents methodologies for organizations experiencing difficulties implementing top-down approaches to build institutions' capacity to adapt to the impact of climate change in developing countries. It also emphasized the benefits of utilizing the hybrid approach to develop CCA initiatives. The research has shown that it is more feasible to transfer to a hybrid approach when encountering difficulties implementing top-down and bottom-up approaches (Matland, 1995; OECD, 2013). The hybrid approach presents an integrated pathway for evaluating and implementing CCA projects by establishing a comprehensive integration of the two approaches. The hybrid approach effectively reduces governments' unnecessary control and inflexibility, centrally managing the development from their headquarters/capitals (Desai & Hulme, 2004; OECD, 2013; Simmons, 2021).

Consequently, developing this framework also contributes substantially to strengthening the capacity and capability of institutions in developing countries, specifically in LDCs and SIDS, to build the resilience of these institutions to adapt to the impact of climate change. The study is situated within the framework of sustainable capital theory focusing specifically on human capital theory (ILO, 2010; Simmons, 2018, 2021; UNDP, 2015). It is framed within the context of enhancing human capital through skills, knowledge and livelihood needs as a mechanism for building the resilience to adapt to climate change (Simmons, 2021).

Methodological Contribution to the Global Body of Knowledge

The new climate change adaptation framework for capacity-building in developing countries is part of the methodological contributions to the global body of knowledge. The development of the conceptual framework on climate change adaptation framework for capacity-building in developing countries will deliver exceptional and appropriate approaches to building the capacity of institutions in developing countries, specifically SIDS and LDCs, to adapt to the impact of climate change. The framework offers policymakers, academics and practitioners an approach to provide institutions and the populations with the knowledge and skills to build the resilience of these organizations to adapt to the impact of climate change.

The framework makes an academic contribution to the global body of knowledge on sustainable livelihoods, capacity development, climate change adaptation, poverty alleviation, leadership development and sustainable development within the context of developing countries. The conceptual framework is relevant for evaluating the impact of climate change on institutions. It is also applicable to achieving sustainable communities in developing countries. This framework can be utilized as a mechanism by policymakers and senior government officials to create policies and programmes to build the capacity of institutions by enhancing their resilience to adapt to the impact of climate change.

The conceptual framework is also valid for further research on vulnerabilities and resilience in ecosystem management, disaster risk management and preparedness, economic development, sustainable livelihoods, sustainable tourism, sustainable agriculture, sustainable fisheries and sustainable forestry ecosystem.

The monitoring and evaluation components are integral elements of the conceptual framework; however, it is extended to include development concepts such as accountability and learning. The accountability elements of the framework are included to indicate the performance of the institutions concerning the CCA-related treaties and agreements signed by developing countries in relation to their governance and financial systems. Learning is related to a) the indigenous technical knowledge, which is present and practised in most developing countries, as well as b) assessing the extent to which the notions of a learning organization are relevant and applicable to building the capacity of institutions to adapt to the impact of climate change (Mulongo, 2012).

6.5.3 Limitations of the Climate Change Adaptation Framework for Capacity-Building in Developing Countries

The framework is valid for challenging problems relating to the impact of climate change on developing countries, specifically in SIDS and LDCs. The research for the book was conducted mainly in institutions and projects in the Caribbean (St Vincent and Barbados), Africa (Sierra Leone and Uganda) and internationally (the Commonwealth). Therefore, there may be difficulties with generalizing the research outcomes due to the differences and variances in the development context of developing countries.

There may be occasions when the climate change adaptation framework for capacity-building in developing countries created in this research would not be appropriate to build the institutional capacity of organizations to adapt to climate change in other developing countries.

Notwithstanding the ambiguity in realizing a comparable conclusion when using the framework across different regional and national contexts in developing countries, researchers are encouraged to implement the conceptual framework in other areas of study to explore whether it is appropriate for other locations and situations. The variations in the context of developing countries would offer exceptional possibilities and opportunities for research to improve the quality of the framework.

The framework may not include a comprehensive body of knowledge that would make it more versatile and adequate to react to and possess the capacity to help all stakeholders desirous of using the conceptual framework. It is essential to note that despite the framework's limitations, it would be necessary to help policymakers, researchers and practitioners find opportunities to improve the management and operation of the institutions to adapt to the impact of climate change in developing countries.

6.6 Further Research

Although academic research on the technical issues of climate change and its impact on natural ecosystems exists, the research on the effects of climate change on human and social systems in developing countries is very limited. The mismatch between the skills and knowledge provided by educational institutions and those that are required by the market has been highlighted in numerous studies as a failure to enhance the capacity of institutions and countries to build the resilience of people to address the climate challenges (Carter, 2008; ILO, 2010, 2013; Simmons, 2021, 2022; UNDP, 2015). However, efforts to mobilize financial and technical resources to support CCA initiatives are severely impacted by the absence of capacity in developing countries to develop and manage CCA projects.

This lack of institutional capacity has resulted in the absence of national policies on climate change in numerous LDCs and SIDS. In a situation where policies exist, the governments are faced with (a) the absence of a viable national development plan, which provides the mechanism for mainstreaming CCA across all development sectors and programmes, and (b) the absence of public and private institutions with responsibility for driving CCA. Therefore, for national climate change policies to be worth the very paper printed on, there must be an ongoing injection of resources to build the capacity of these institutions to ensure appropriate plans and programmes are implemented. The strategy should be supported by ongoing research to assess (a) the effectiveness of climate change adaptation plans and programmes implemented and (b) the capacity-building initiative established by institutions to develop the adaptive capacity and resilience to adapt to the impact of climate change. To bridge the data gap in capacity-building in institutions and climate change adaptation, researchers would need to be engaged to implement further work in areas such as climate change impacts, vulnerability and adaptation and the impact of climate change on specific sectors of the population in developing countries.

Research has shown that sea-level rise (SLR), hurricanes, temperature rise and other climate stressors will negatively impact developing countries (UNDP, 2015; Nurse, 2014; IPCC 2014a, b, c, 2022). According to Simpson et al. (2010), SIDS would continue to experience the worst rising sea level. Despite the numerous research conducted, there are still gaps in the body of data on the impact of climate stressors on population and ecosystems in developing countries. Therefore, there is a need for further research to assess the impact of climate stressors on specific population sectors and their institutions.

6.7 Lessons Learned and Good Practices

This section of the chapter presented a summary of the lesson learned and good practices and made recommendations to enhance the capacity-building programmes to build people's resilience and their institutions to adapt to climate change and development calamities in developing countries.

(a) Lesson Learned from the Case Studies.

The lesson learned evolving from the case studies on capacity-building:

- JEMS strategy is a bottom-up approach focused on putting people first by ensuring that they are at the centre of the development process (Leach, 1996; Angelson, 2011; Commodore et al., 2017; Eicken et al., 2021). The JEMS model has demonstrated over the past 45 years that the bottom-up participatory development approaches possess the potential to build community cohesion, foster empowerment and enhance sustainability, whereby people take responsibility for shaping and determining their own destiny.
- While migration has taken its toll on the development of the communities and the country, JEMS continues to maintain a cadre of over 1000 volunteers on whom it can call at any time to service the organizations and participate in community development programmes. Maintaining such a vital base is critical to the organization's longevity and ability to remain a constant and effective agent for change in the community. As people continue to migrate to seek "greener pastures" overseas, JEMS continue to find ways of maintaining committed membership and leadership; hence continuing work on enhancing its capacity-building programme is a noteworthy pathway for the organization (Enekmark, 2003; Mizan Khan et al., 2019).
- The implementation of popular capacity-building participatory methodologies such as cultural wake and community mapping contributed to the involvement and participation of people in the national development planning processes. These strategies give the people a sense of ownership, commitment and stake in the overall development of their community and countries.
- Educating the people of the communities on the development challenges confronting the community and the nation as a whole would enhance their participation in their country's overall development. The process empowers people to take leadership and other critical development roles in the governance of their communities/country and can lead to achieving Sustainable Development Goals.
- YEEP has contributed enormously to enhancing the capacity of its implementing partners, namely, NAYCOM, MoYA, NGOs, universities, colleges and TVET centres participating in the project. The staff of these agencies were involved in the training and consultation process and acquired new knowledge, skills and competencies, which contributed to the

overall management and operation of the education and development institutions and projects.

- The capacity-building methodologies enhance the technical and institutional capacity of the institutions to better monitor, evaluate, report, train, track and manage their youth development programme in Sierra Leone.
- The capacity-building programmes implemented by CYP reveal improvements in its management and operation and enhanced better-quality assurance mechanism processes. The improvements in the quality of CYP's capacity-building work resulted in the organization attracting new partners to support its programmes in member states (e.g. partnership with The Body Shop and Commonwealth Bank in the Pacific region).
- CYP's work creates opportunities for young people to participate in developing policies, governance and decision-making through the National Youth Policy development programme. It dedicated its work to enabling young people to participate in policymaking and making decisions on the development they desire to see implemented locally.
- CYEN observed that increases in awareness and education do not lead to actions and change in the environment and climate change adaptation. There is urgency on the part of the organization to re-examine the strategies, approaches, programmes implemented and the outcomes anticipated to ensure that they are in line with the goals of the organization. This approach would help the organization build young people's resilience to adapt to the impact of climate change and development calamities and lead to achieving Sustainable Development Goals (SDGs).
- Implementing refresher/orientation training courses and workshops contributes to building the capacity of teachers and instructors to deliver quality education provisions for the RAY in Northern Uganda.
- The involvement of technical educators from the local government authorities in Northern Uganda to support teachers was essential to raising the credibility and awareness of the non-formal education component of the RAY and providing a long-term base of support for the continued development of the project.

(b) Best Practices Evolving from the Case Studies on Capacity-Building.

A summary of the best practices evolving from the case studies:

- The mobilization of skilled volunteer workers by JEMS to undertake work to rehabilitate the housing and infrastructures in villages in the southeast of main St Vincent and the Grenadines that were destroyed by storms and hurricanes contributed to reducing the cost of rehabilitation of houses and other community facilities. Moreover, the initiative reduced mistrust between government workers and the community residents.
- Women and youths were deemed the most vulnerable sector of society; however, they were immediately mobilized and contributed to the development of their community. The participation of women in the projects attracted men to the volunteering project.
- The utilization of participatory practices was viewed as a key strategy which enhances the participation of the poor, poverty-stricken and marginalized to transform squatting settlements into fully independent communities with basic infrastructure, including water, paved roads, sanitary facilities and community education programmes.
- The aided help development approach is viewed as an effective community development strategy in St Vincent and the Grenadines. It combines technical assistance and material resources from the government and matches with people's labour and food to enhance the basic infrastructure of communities.
- The increase in the number of young people desirous of accessing the Business Development Services programmes in Sierra Leone was attributed to the values placed on the high

quality of the training programme and internships provided to young people by businesses development agencies.

- Provision of incentives to poor young men and women, such as transportation, lunch and equipment packages by the YEEP, enables poor young men and women to benefit from such development projects, which would not typically be provided with the opportunity to participate in such projects.
- CYP has instituted the hybrid approach, utilizing the strengths of top-down and bottom-up approaches to implement its capacity-building programmes in member countries. It also uses participatory methodologies to complement the delivery of its training, internship and coaching programme.
- CYEN takes on a holistic, balanced and cooperative approach to tackle environmental and sustainable development challenges in the Caribbean region. When the people deemed the end users of the natural resource base of a country are elevated to be part of the emerging management model implemented by the organization to resolve the environmental challenges in the Caribbean, then the project would be a success for all stakeholders.
- The bottom-up approach utilized by CYEN is viable for empowering and mobilizing young people and their organizations/communities to participate in environmental and sustainable development initiatives. The approach values the beneficiaries' contribution to achieving the organization's outcomes at community, national and regional levels.
- The participatory approach is used by CYEN to build capacity and enhance the quality of the project management process. It empowers young people and other stakeholders to participate in the project development by building partnerships among key stakeholders.

6.8 Recommendations for Future

The recommendation for enhancing capacity-building in institutions in the future:

- Investing in capacity-building to enhance conflict resolution and mediation skills is a worthwhile initiative which could reduce the apparent partisan perceptions often associated when non-governmental organizations and governments are at variance on issues of sustainable development. In fact, it is a well-established fact that JEMS' collaborative linkages with its numerous partners and stakeholders have been pivotal to the organization's growth, development and sustenance. One of its key strengths is its collaborative solid links and cooperation among stakeholders and partners. The organization is urged to continue working on this future development trajectory.
- The CDD should continue its efforts to inject funding and technical resources to develop training and other capacity-building programmes in communities to enhance communities' awareness and capability to develop and implement programmes to resolve the development challenges in communities.
- Continue the work of CDD on building partnerships with internal and external agencies to mobilize financial resources to build the capacity of staffs and partner agencies to implement viable, sustainable programmes at the community and national levels.
- Development of vital staff appraisal systems backed up with promotions and recognition of staffs and volunteers that work hard; appreciate their efforts openly so that they feel recognized, and then reward them with the right incentives to keep their enthusiasm going.
- The NAYCOM is viewed as a collection of human beings working together, so building core relationship skills, e.g. the ability to collaborate and communicate, is one of the most important initiatives it should encourage.
- The CYP should continue to collaborate with UNICEF, UNFPA, WHO, universities and colleges to explore opportunities, including avenues for mobilizing financial resources to build the capacity of member governments, universities and youth networkers to develop employment and sustainable livelihoods, climate change adaptation, youth policy, youth

work training and governance initiatives. Due to the gap in data resources, the organization should invest in research on developing best practices and strategies/approaches/processes which could be adopted and replicated throughout the Commonwealth.

- Governments in developing countries should support policy initiatives to include young people in the delegations attending international meetings, including the Commonwealth Heads of Government Meeting, UNFCCC and other high-level meetings.
- Mobilization of adequate finances to support the implementation of its programmes is necessary for CYEN. The resources should be used to fund environmental and sustainable development programmes. It should consider lobbying governments in the region to offer subventions to cover its administrative and operational costs.
- To enhance the capacity of CYEN to mobilize young people to implement viable climate change and sustainable development projects at the national level, governments should explore the possibility of (a) providing office space for the National Youth Network (NYN) to function and (b) seconding staff from the public service to provide technical assistance and support to NYN.

References

Adger, W. (2003). Social capital, collective action, and adaptation to climate change. *Economic Geography, 79*(4), 387–404. Available at: http://link.springer.com/chapter/10.1007/978-353192258-4_19

Angelsen, A., Larsen, H. O., Lund, J. F., Smith-Hall, C., & Wunder, S. (2011). *Measuring livelihoods and environmental dependence*. Methods for Research and Fieldwork.

Armitage, D., & Plummer, R. (Eds.). (2010). *Adaptive capacity and environmental governance*. Springer-Verlag.

Barnes, J., Simmons, A., Abola, C., Nkawanzi, V., Semay, S., & Greeley, S. (2013). *Norwegian refugee council recovery of Acholi youth independent evaluation report*. Impact Ready; Version 2.4: 25th November 2013.

Bloomfield, G., Bucht, K., Martínez-Hernández, J. C., Ramírez-Soto, A. F., Cheseña-Hernández, I., Lucio-Palacio, C. R., & Inzunza, E. R. (2017). Capacity-building to advance the United Nations sustainable development goals: An overview of tools and approaches related to sustainable land management. *Journal of Sustainable Forestry, 37*, 157. https://doi.org/10.1080/10549811.2017.1359097

Butler, J. R. A., Wise, R. M., Skewes, T. D., Bohensky, E. L., Peterson, N., Suadnya, W., Yanuartati, Y., Handayani, T., Habibi, P., Puspadi, K., Bou, N., Vaghelo, D., & Rochester, W. (2015). Integrating top-down and bottom-up adaptation planning to build adaptive capacity: A structured learning approach. *Coastal Management, 43*, 346–364., Taylor & Francis Group, LLC; ISSN: 0892-0753 print/1521-0421 online. https://doi.org/10.1080/08920753.2015.1046802

Callaghan, I. (2015). *Climate finance after COP21: Pathway to effective financing of commitments and needs: Invest watch.* Ian Callaghan Associates Limited 2015 Final 011215_MG2.

Carter, R. (2008). *Caribbean youth*. An Integrated Literature Review.

Chamber, R. (2002). *Participatory workshops: A sourcebook of 21 sets of ideas and activities* (Vol. 794, pp. 421–434). Earthscan.

Chanika Mataya, D., Vincent, K., & Dougill, A. J. (2020). How can we effectively build capacity to adapt to climate change? Insights from Malawi, climate and development. *Francis and Taylor group, 12*(9), 781–790. https://doi.org/10.1080/17565529.2019.1694480

Charles, H. (2006). The commonwealth response to global challenges facing the 21st century: Commonwealth youth and development. *Department of Commonwealth Studies; UNISA (University of South Africa Press), 4*(2), 49.

Chatenoux, M., & Wolf, A. (2013). *Ecosystem-based approaches for CCA in Caribbean SIDS* (p. 64). UNEP/GRID-Geneva and ZMT Leibniz Centre for Tropical Marine Biology.

Chevallier, R. (2010). Integrating adaptation into development strategies: The southern African perspective. *Climate and Development, 2*(2), 191–200. https://doi.org/10.3763/cdev.2010.0039

Commodore, A., Wilson, S., Muhammad, O., Svendsen, E., & Pearce, J. (2017). Community-based participatory research for the study of air pollution: A review of motivations, approaches, and outcomes. *Environmental Monitoring and Assessment, 189*, art. 378.

Dany, V., Bowens, K. J., & Miller, F. (2015). Assessing the institutional capacity to adapt to climate change: a case study in the Cambodian health and water sectors. *Climate Policy, 15*(3), 388–409.

Dessai, S., & Hulme, M. (2004). Does CCA policy need probabilities? *Climate Policy., 4*, 107–128.

Eicken, H., Danielsen, F., Sam, J. M., Fidel, M., Johnson, N., Poulsen, M. K., Lee, O. A., Spellman, K. V., Iversen, L., Pulsifer, P., & Enghoff, M. (2021). Connecting top-down and bottom-up approaches in environmental observing, *BioScience, 71*(5), 467–483..

Enemark, S. (2003). *Understanding the concept of capacity-building and the nature of land administration systems*. FIG Working Week, April.

Hernandez-Delgado, E. A. (2015). *The emerging threats of climate change on tropical coastal ecosystem service, public health, local economies and livelihoods sustainability of small islands: Cumulative impacts and synergies*. Marine Pollution Bulletin, University of Puerto Rico, Centre for Applied Tropical Ecology and Conservation, Coral Reef Research Group Puerto Rico, USA.

Horstmann Britta and Gunther Schulz-Heiss. (2014). *Providing international adaptation finance for vulnerable communities: A study on potentials and limits of social investment funds, The German Development Institute* (pp. 1860–0468). ISSN.

ILO. (2010). *A skilled workforce for sustainable growth: A G20 training strategy* (p. 2010). International Labour Office.

ILO. (2013). Sustainable development, decent work and green jobs. In *International labour conference, 102nd session*. International Labour Office.

IPCC. (2014a). Climate change (2014): Synthesis report. In *Contribution of working groups I, II and III to the fifth Assessment Report of the Intergovernmental Panel on climate change* (p. 151). IPCC Geneva.

IPCC. (2014b). Summary for policymakers. In C. B. Barros, V. R. Dokken, D. Mach, K. J. Mastrandrea, M. D. Bilir, T. E. Chatterjee, M. Ebi, & K. L. Estrada (Eds.), *Climate change impacts, adaptation and vulnerability*. Contributions of the Working Group II to the Fifth Assessment Report, Field.

IPCC. (2014c). Climate change: Synthesis report. In *Contribution of working group 1, 11, 111 to the fifth assessment report of the intergovernmental panel on climate change* (p. 151). IPCC Geneva.

IPCC. (2022). *Climate change 2022 mitigation of climate change, summary for policy makers: Working group 111 contribution to the sixth assessment report of the international panel on climate change (AR6 WG 111)*. IPCC.

Jaja, J., Dawson, J., & Gaudet, J. (2016). *Using social network analysis to examine the role that institutional integration plays in community-based adaptive capacity to climate change in Caribbean small Island communities* (Vol. 22, p. 424). Department of Geography, Environment and Geomatic, University of Ottawa: Routledge, Local Environment.

Jayatilaka, W. (2003). Capacity-building for sustainable development. *Journal of the National Science Foundation of Sri Lanka, 31*, 81–97.

Leach, M. (1996). *A revolution of ideas and action: Interview with Andrew Simmons*. Share International: July/August 1998 Issues of Share International. https://www.share-international.org/archieves/environmental/en_mlrevolution.htm (download 09/05/2022)

Matland, R. (1995). Synthesising the implementation literature: The ambiguity-conflict model of policy implementation. *Journal of Public Administration Research and Theory, 5*(2), 145–174.

Mizan, K., Mfitumukiza, D., & Huq, S. (2019). Capacity-building for implementation of nationally determined contributions under the Paris agreement. *Climate Policy., 20*, 499. https://doi.org/10.1080/14693062.2019.1675577

Mulongo, G. (2012). *Human Capital in Education: Principles, critiques and current thinking*. Institute of Education, University of London.

Nurse, L. A., McLean, R. F., Agard, J., Briguglio, L. P., Duvat-Magnan, V., Pelesikoti, N., Tompkins, E., & Webb, A. (2014). Small islands. In V. R. Barros, C. B. Field, D. J. Dokken, M. D. Mastrandrea, K. J. Mach, T. E. Bilir, M. Chatterjee, K. L. Ebi, Y. O. Estrada, R. C. Genova, B. Girma, E. S. Kissel, A. N. Levy, S. MacCracken, P. R. Mastrandrea, & L. L. White (Eds.), *Climate change 2014: Impacts, Adaptation, and Vulnerability* (pp. 1613–1654). Part B: Regional Aspects. Contribution of Working Group II to the Fifth Assessment Report of the Intergovernmental Panel on Climate change Cambridge University Press.

OECD. (2013). *The nature of policy change and implementation: A review of different theoretical approaches*. Organisation for Economic Cooperation and Development.

Ojwang, L., Rosendo, S., Celliers, L., Obura, D., Muiti, A., Kamula, J., & Mwangi, M. (2017). Assessment of coastal governance for climate change adaptation in Kenya. *Earth's Future, 5*, 1119–1132. https://doi.org/10.1002/2017EF000595. Received 14 APR 2017Accepted 6 OCT 2017Accepted article online 16 OCT 2017Published online 14 NOV 2017© 2017.

Organisation for Economic Co-operation and Development (OECD). (2006). *The challenge of capacity development: Working towards good practice* (DAC Guidelines and Reference Series) (p. 46). OECD.

Pahl-Wostl, C., Conca, K., Kramer, A., Maestu, J., & Schmidt, F. (2013). Missing links in global water governance: A processes-oriented analysis. *Ecology and Society, 18*(2), 33. https://doi.org/10.5751/ES-05554-180233

Poschen, P. (2015). *Decent work, green jobs and the sustainable economy: Solution for climate change development*. International Labour Organisation.

Simmons, A. (2006). Youth development index. *Commonwealth Journal on Youth and Development, UNISA, South Africa, 4*(2), 100–113.

Simmons, A. (2018). *Impact of climate change on youth in small Island communities: The case of St. Vincent and the Grenadines* (This thesis is submitted to De Montfort University in partial fulfilment of the requirements for the award of Doctor of Philosophy). School of Engineering and Sustainable Development De Montfort University.

Simmons, A. (2021). *Impact of climate change on youth in small Island developing communities* (Sustainable Development Series). Springer.

Simmons, A. (2022). *Climate change adaptation framework and youth entrepreneurship in West Africa* (Sustainable Development Goals Series). Springer.

Simpson, M. C., Scott, D., Harrison, M., Simm, R., Silver, N., O'Keeffe, E., Harrison, S., Taylor, M., Lizcano, G., Rutty, M., Stager, H., Oldham, J., Wilson, M., New, M., Clarke, J., Day, O. J., Fields, N., Georges, J., Waithe, R., & McSharry, P. (2010). *Quantification and magnitude of losses and damages resulting from the impact of climate change: Modelling the transformational impacts and costs of sea level rise in the Caribbean*. West indies: United National Development Programme (UNDP).

Smit, B., & Wandel, J. (2006). Adaptation, adaptive capacity, and vulnerability. *Glob Environ Change, 16*, 282–292. https://doi.org/10.1016/j.gloenvcha.2006.03.008

Stachowiak, S., Robles, L., Habtermanion, E., & Maltry, M. (2016). *Beyond win: Pathway for policy implementation*. ORS Impact/The Atlas Learning Project.

Talawira, I., & Simmons, A. (2004). *Youth development at the Centre of the Commonwealth Agenda*. Commonwealth Youth Programme (CYP): Forum 21: European Journal on Youth Policy No3 12/2004.

UNDP. (2015). *Human development report*. Work for Human Development, United National Development Programme (UNDP).

United Nations (UN). (2015). *Resolution adopted by the General Assembly on 25 September 2015. A/RES/70/1 Transforming our world*. The 2030 Agenda for Sustainable Development.

University of the West Indies. (2011). *The National Youth Mainstreaming Strategy Workshop*. Presentation of Strategy Document June 29–30, 2011: The Centre for Leadership and Governance, The University of the West Indies; The Ministry of Youth, Sports & Culture, The National; Centre for Youth Development and the International Development Bank.

Appendix: Questions for Key Informants (These Were Administered as a Questionnaire or as Interviews)

Research: Building Institutions for Resilience: Combatting Climate Change
Researchers: Dr Andrew Simmons and Andree Ann Simmons

Please fill in the information below and send it back to me at [email]

Name of official:
____
____

Job title:
____
____

Age range (Please circle): 20–24 25–29 30–34 35–39 40+

Address:
____
____
____
____
____
____

Contact: Phone: ________________ Email:
____

Name(s) of institution/organization:
____
____

Please circle which organization best suits where you work:
(a) Government
(b) Civil society (NGO)
(c) None of the above (Please describe your organization):
____
____
____

How is the organization funded?
____
____

What are the main programmes implemented by the organization?
____
____

Who are the beneficiaries of the programmes?
____
____

A. Simmons, A.-A. Simmons, *Building Institutions for Resilience*, Sustainable Development Goals Series, https://doi.org/10.1007/978-3-031-28881-4

Please respond to the following questions.

1. Tell me about yourself and the role(s) you play in the organization/institution?

2. What are the **strengths** of your organization's programmes, projects and strategies?

3. What are the **weaknesses** of your organization's programmes, projects and strategies?

4. What **challenges** hinder your organization from accomplishing its objectives?

5. What did you and your team do to **resolve** the challenges encountered by the organization?

6. What are the **organization's capacity development problems (human)**?

7. How has the organization worked to resolve its human capacity development problems?

8. What **methodologies** are used by the organization to train persons to develop the overall capacity of partner organizations and stakeholders?

9. How effective are the **methods** used by the organization to train persons and develop the overall capacity of partner organizations and stakeholders?

10. What were the key **lessons from implementing the organization's capacity development programmes/projects to secure** a more sustainable future?

11. What do you see as **best practices** you can extract from your organization's work to enhance its human capacity to develop beneficiaries and partner organizations in the future?

12. What **recommendations** would you consider for improving the human capacity of the organization to develop beneficiaries and partner organizations in the future?

13. What could have been done to **up-scale the** organization's capacity to adequately respond to the community/country/region needs?

Thank you for your responses.

Glossary

Adaptive capacity the ability of a system to adjust to climate change, including climate variability and/or extremes, to moderate potential damages, cope with consequences and take advantage of the opportunities climate change may bring.

Bottom-up approach to CCA characterized as observing or monitoring efforts defined and undertaken at the local scale and brought forward to higher-level bodies, often focusing on supporting outcomes desired by a local community. Bottom-up approaches originate at the local community level, with observing requirements tied to management outcomes that community members or institutions are seeking to achieve.

Capacity-building endeavours and actions to build individuals' and organizations' understanding, skills, and knowledge base to achieve their development goals. It is also described by Gordon and Chadwick as activities to build the awareness, skills and knowledge base of individuals and organizations.

Complexity leadership theory (CLT) an integrative framework of leadership functions, including political-administrative, connective, adaptive, enabling and dissemination functions. Its shared purpose is to enhance the adaptive capacity of governance networks.

Climate change a change of climate which is attributed directly or indirectly to human activity that alters the composition of the global atmosphere and which is in addition to natural climate variability observed over a comparable time.

Climate change Adaptation refers to human actions to build resilience and reduce vulnerability to the existing impact of climate change, which is closely aligned with the concept of disaster risk reduction.

Climate change Education for Sustainable Development (CCESD) focuses on "preparing people from all walks of life to plan for, cope with and find solutions for issues that threaten the sustainability of our planet".

Climate Justice defined as the capacity of social, economic and ecosystems to cope with a hazardous event or trend or disturbance, responding or reorganizing in ways that maintain their essential function, identity and structure as well as biodiversity in the case of ecosystems while also maintaining the capacity for adaptation, learning and transformation.

Empowerment creating and supporting the enabling conditions under which young people can act on their behalf, and on their terms, rather than at the direction of others.

Exposure the presence of people, livelihoods, species or ecosystems, environmental functions, services, infrastructure or economies and social or cultural assets in place and settings that could be adversely affected.

Green Growth Growth that is efficient in its use of natural resources, clean in that it minimizes pollution and environmental impacts and resilient in that it accounts for natural hazards and the role of environmental management and natural capital in preventing physical disasters…this growth needs to be inclusive.

A. Simmons, A.-A. Simmons, *Building Institutions for Resilience*, Sustainable Development Goals Series, https://doi.org/10.1007/978-3-031-28881-4

Hybrid approach to CCA incorporating or linking the strengths of top-down and bottom-up approaches, i.e. combing the micro-level variables of bottom-up and the macro-level variables of the top-down approaches in the development and implementation of climate adaptation projects to benefit from the advantages and strengths of both systems by enabling the various levels of stakeholders to interact and influence project outcomes. Combining both approaches highlight their strengths and minimizes their weaknesses.

Human capital the skills, knowledge and ability to labour and good health together enable people to pursue different livelihood strategies and achieve their livelihood objectives.

Human Capital Models presume that an investment in education is made with the expectancy of positive returns to the individual or state on their investment.

Human Capital Theory the theoretical framework responsible for adapting education and development policies.

Leadership inspiring individuals to give their best to achieve the desired result. He contends that it is about getting people to move in the right direction, gaining commitment and motivating them to achieve their goals.

Life learning approach emphasizes the need for policymakers, education and training institutions to link training and skills provided with market needs as an initial strategy to prepare youth to make full use of future opportunities created by climate change.

Livelihood comprises the capabilities, assets and activities required for a living.

Gender Mainstreaming a strategy to improve the quality of public policy programmes and projects to ensure a more efficient allocation of resources, i.e. better results mean increased well-being for both women and men and the creation of a more socially just and sustainable society.

Resilience the ability of human or ecological systems to cope with and adapt to environmental changes.

Sample the subgroup of a target population studied, i.e. the group from which data are collected.

Sensitivity the degree to which a system is affected by or is responsive to climate inducements, i.e. it is also referred to as the degree to which a structure or organism is affected, either unfavourably or beneficially, by climate-related impacts.

Shocks destructive events that occur unexpectedly and last for a concise period, such as tsunamis, earthquakes, storms, volcanic eruptions, landslides, avalanches and wildfires.

Small Island Developing States (SIDS) recognized as a distinct, diverse group of developing countries encountering specific social, economic and environmental vulnerabilities at the United Nations Conference on Environment and Development (UNCED in Rio in 1992), also known as the Earth Summit, held in Rio de Janeiro, Brazil (3–14 June 1992). http://unohrlls.org/about-sids/

Stressors events that happen slowly, with their impacts being felt for longer periods.

Sustainable development "development that meets the needs of the present without compromising the ability of future generations to meet their need".

Sustainable Livelihoods a livelihood is sustainable when it can cope with and recover from the stresses and shocks and maintain or enhance its capabilities and assets both now and in the future without undermining the natural resource base.

SWOT analysis an analytical tool that identifies the internal strengths and weaknesses, as well as the external opportunities and threats of an organization or project. It is used to analyse the data collected for the case studies.

Target population the group being studied, i.e. the group about which one wishes to draw data conclusions.

Top-down approach to CCA government-level decision-making based on long-term regional climate change modelling, impact projections and adaptation strategies identified through technocratic cost-benefit analyses. Top-down CCA is more applicable to government agencies.

Vulnerability the degree to which geophysical, biological and socio-economic systems are susceptible to and unable to cope with adverse impacts of climate change; i.e. it is the degree to which a system is susceptible to injury, damage or harm.

Youth empowerment Youth are empowered when they acknowledge that they have or can create choices in life, are aware of the implications of those choices, make informed decisions freely, take action based on that decision and accept responsibility for the consequences of those actions.

Youth the UN defines young people as a cohort of persons from 15 to 24 years of age, whereas in the Commonwealth, youth is defined as persons aged 15–29 years. Qualitatively, youth is defined as a phase when a person moves from a time of dependency (childhood) to independence (adulthood).

Index

A. Simmons, A.-A. Simmons, *Building Institutions for Resilience*, Sustainable Development Goals Series, https://doi.org/10.1007/978-3-031-28881-4

I

L

N

P

R

S

T

W

Y

The manufacturer's authorised representative in the EU is Springer Nature Customer Service Centre GmbH, Europaplatz 3, 69115 Heidelberg, Germany. If you have any concerns regarding our products, please contact ProductSafety@springernature.com

Printed and bound by CPI Group (UK) Ltd, Croydon, CR0 4YY
28/09/2026
02224071-0020